"I have had the pleasure to work with and observe Rocky's work over the last decade. Rocky has always understood the deep impact that technology and innovation has on industries, and the way we live and work. He has the ability to translate technological change into insights for business and community leaders. Rocky is both a visionary and a change agent. I have always made time to read his articles and publications."

David Thodey – Chairman CSIRO

"Rocky is a rare breed of futurologist who can combine a very strong data-driven approach, with an almost intuitive ability to be able to see things that others can't, that in many cases are right in front of them. Combine that with a powerful communication style and desire to genuinely help others, and you have the perfect storm for being able to accurately predict and ultimately respond quickly to a very fast-changing environment. As someone who is obsessed with contingency planning and crisis management, this is a must-read for anyone who wants to be prepared for the decade ahead. His insights and predictions for the next 10 years have never been more important."

Kerwin Rae – Chief Executive Officer

"This report is an important contribution to our understanding of the impetus to reach net-zero as a digital economy and the enormous benefits that may be yielded for businesses and society if we get this right. From this seminal piece of work, technology leaders stand to gain a deep appreciation of their role in making these targets a reality, with innovation and connectivity providing both the engine and backbone in the race to decarbonise."

Simon Bush – Chief Executive Officer Australian Information Industry Association

"I have known Rocky a long time and whether I am listening to his speeches or reading his research and predictions I always feel confronted in a good way. Whether you agree with his views or not, he makes you think. And think differently!"

Alexis George – Chief Executive Officer AMP

"Rocky has a deep, globally based understanding of the zeitgeist of the moment. He knows what to look for when separating important events from noise. He has an inclusive approach to data gathering and an understanding of current and potential changes in the work environment, especially as they relate to Industry 4.0 and its transformative role they can play."

Michelle Blum – Chief Executive Officer Australia-Israel Chamber of Commerce

"Scopelliti's fascinating *Australia 2030 Research*© takes the pulse of current reality, cutting through the maze of political rhetoric, popular assumptions, and glossy opinions to reveal the blinkers off reality of the past and present, exposing the risks and opportunities facing Australia. But the report doesn't just expose and explore the problem domain, it goes on to deliver a well-researched roadmap to move conversations and decisions forward on the most pressing issues that both unite and divide opinion. Succinctly offering decision makers the best possible start at rethinking how Australia can emerge to play a leading role in the region".

"Challenging the status quo throughout, Scopelliti's report comes at a pivotal time not only in Australian history but that of the world as climate change reshapes nations, industries, businesses, and governments and pushes technology along new frontiers. Will Australia seize the moment and enter the new decade leading the region in terms of climate change? If you think this question can only be answered by politicians, you'd be wrong. This report shows you why".

Tremaine du Preez, PhD, Decision Scientist, and author o
DECIDE: The Art and Science of Choosing Wisely and Ra
Thinkers – Preparing your Child for the Journey of Life

Published in Australia by

FUTUROLOGY
MEDIA

Rocky Scopelliti
rocky@rockyscopelliti.com
www.rockyscopelliti.com

First published in Australia 2022

National Library of Australia Cataloguing in Publication entry

NATIONAL LIBRARY OF AUSTRALIA

A catalogue record for this book is available from the National Library of Australia

ISBN: 978-0-6456630-3-7 (paperback)
ISBN: 978-0-6456630-1-3 (e-book)

Cover design by Studio Franzini
Layout and typesetting by Sophie White
This report features images by vecteezy.com and unsplash.com

Printed by Ingram Spark

AUSTRALIA'S TECHNOLOGICAL RACE TO NET ZERO CARBON EMISSIONS

How can Australia increase its capacity to adapt to a world in accelerated Climate Change?

About the *Australia 2030 Research*©

In this report, I have brought together new insights from the ***Australia 2030 Research***© longitudinal program I commenced in 2020. This research investigates Australian board directors', chief executives', and professionals' attitudes toward the decade ahead. The results of that first study were published in my last book titled *'Australia 2030 – Where the Bloody Hell Are We?'*

The focus of this second round of *Australia 2030 Research*© is on climate change. To date, it is the largest national study of its type investigating the opinions of Australian board directors, chief executive officers, technology leaders, c-suite leaders, and professionals. The material describes the readiness of their organisations, the challenges and roadblocks faced, and what needs to be done to meet our new environmental carbon emissions targets.

To all of them, I am sincerely grateful for your participation, encouragement, and insights.

In a nutshell, the primary concerns board directors and chief executives expressed about their organisations' environmental strategic readiness over the coming decade are:

1. Clarity of benefits and unclear returns on investment (ROI).
2. Lack of skills and reskilling.
3. Under investment in the necessary enabling technologies.

The *Australia 2030 Research*© further showed that environmental programs are both data and technology intensive with diverse and highly fragmented sources requiring complex transformations.

An important finding is that to exponentially transform Australia's industries and organisations strategically and operationally, the Chief Information Officer role is critical.

Chief Information Officers are the enablers and systems leaders of their organisation's environmental strategy and initiatives.

In the following sections I capture the status quo as outlined by Australia's industrial and business leaders and explore the power of digital technologies to radically transform them.

The report will equip board directors, leaders, investors, policy makers and professionals with the technological insights to better understand what they need to do to mitigate the impact on the risks and explore the opportunities inherent in decarbonisation. It includes resetting strategies, investment, resources, capital allocation, and operational execution to respond to climate change.

I welcome the opportunity to further present in-depth briefings to you and your leaders on what these insights mean, and how they can prepare your organisation for a decarbonised 4th Industrial Revolution world.

Let's begin the systems-led, exponentially paced, and technological-enabled journey to zero carbon emissions.

ROCKY SCOPELLITI

The Second *Australia 2030 Research*© Scope

This, the second program in the *Australia 2030 Research*© series, investigates specifically how ready Australian industries, organisation and leaders are for climate change and the new environmental landscape. By invitation, 539 Australian leaders and professionals (183 of whom classified themselves as board directors or chief executives), were invited to participate in the quantitative and qualitative research conducted online over September/October 2022.

***Australia 2030 Research*® 2022 Sample Size**

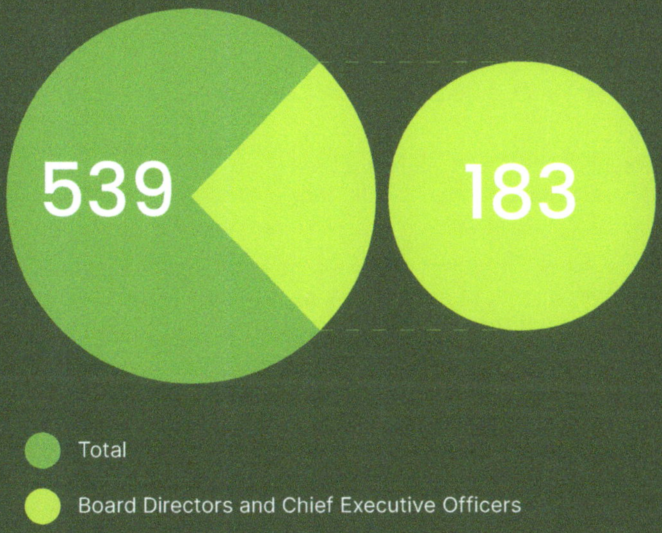

Australia 2030 Research© Questions

- Which issues do you feel will affect Australia over the next 10 years?
- Which strategic priorities will be the most critical for your organisation over the coming 10 years?
- Does your organisation have environmental strategy, policies, and programs?
- If so, how mature are your organisation's environmental strategies, policies, and programs?
- Do you integrate environmental considerations into strategic decision making and operational execution?
- Has your organisation established a robust environmental goal-setting process that considers stakeholder sentiment, business opportunity and corporate capability?
- What role does the Chief Information/Technology Officer or equivalent play in supporting your organisations environmental program(s)?
- What stakeholder groups does your organisation seek to influence by acting on/investing in environmental initiatives?
- Which greenhouse gas (GHG) emissions reduction targets have been set by your organisation?
- Does your organisation's financial planning include the possibility of the future costs associated with carbon emission by way of some form of carbon taxation/pricing scheme?
- How many years from now do you anticipate being forced to pay for carbon (other than through voluntary offsets and investments)?
- What benefits have already been achieved from your organisation's environmental program(s) in 2022?
- What technologies is your organisation investing in for its environmental program(s)?
- What concerns do you have about your organisations environmental strategic readiness for the coming decade?
- Does your organisation measure carbon emissions across enterprise technology?
- How do you feel about the speed of technological and scientific change required to support Australia's environmental goals over the coming decade?
- Which technological or scientific developments do you feel will have the greatest impact on environmental innovation in Australia in the coming decade?
- How do you feel about the role of technological and scientific developments to support climate change.

About the Author

Rocky Scopelliti is a world-renowned futurologist whose purpose is to *'create confidence in our future'.* His pioneering behavioural economics research on the confluence of demographic change and digital technology, has influenced the way we think about our social, cultural, economic, and technological future.

His current book ***Australia 2030! – Where the bloody hell are we?*** is the first major research into Australian professionals' attitudes toward the decade ahead conducted before and during the eye of the COVID-19 storm. His other book *Youthquake 4.0 – A Whole Generation and the New Industrial Revolution'* was launched globally and is now published in Chinese, Vietnamese, Indonesian and Korean languages.

As a media commentator, his unique insights have featured on SKY Business News, The Australian Financial Review, ABC Radio National, The Economist, Forbes, and Bloomberg. As an international keynote speaker, his presentations have captivated audiences across Asia Pacific, the USA and Europe including Mobile World Congress. As a thought-leader, more than 150 boards and leadership teams, including Fortune 100 corporations, each year seek his advice on strategy, emerging technologies, and their future impact.

A distinguished author, his 18 thought leadership research publications have become internationally recognised for their influence including by the World Economic Forum.

In an executive capacity, he is a member of the Optus Enterprise & Business leadership team as a Chief Scientist for Government where he leads the creation of world class research thought leadership to guide the Fourth Industrial Revolution.

In a non-executive capacity, he is an Adjunct Professor (Industry) at the University of Technology Business School, a Non-Executive Director on the board of Community First Credit Union, on the technology advisory board for REST Super and Wake by Reach and is a former board member of the Australian Payments Council.

Educated in Australia and trained in the USA at Sydney and Stanford Universities respectively, he has a Graduate Diploma in Corporate Management and an MBA. He is also a graduate and member of the Australian Institute of Company Directors.

Contents

Foreword

"The countries that build great zero-carbon companies and industries will be the ones that lead the global economy in the coming decades," Gates writes. "Whoever makes big energy breakthroughs and shows they can work on a global scale, and be affordable, will find many willing customers in emerging economies."

Bill Gates
How to avoid a climate disaster: The solutions we have and the breakthroughs we need

Growing our way to net zero carbon emissions

Decarbonising the world's economies, industries, organisations, and populations will be as powerfully disruptive as climate change itself. Policy and technological changes that seek to limit warming and reduce its physical impacts, are already causing disruption to businesses, supply chains and markets.

As with any form of disruption, climate change is creating risks and opportunities for business and governments. The reciprocal, and inextricably linked relationship between climate change and decarbonisation is a top priority for governments, industries, investors, businesses, and leaders world-wide.

For Australia there are many climate change questions to consider. Here are seven;

1. How will state and federal government policy settings change to attract and incentivise global, regional, and local investment?
2. What (additional) Environmental Social Governance (ESG) factors will investors demand in reward for their risks in pursuit of the business opportunities ahead?
3. What product innovations will industries and organisations invest in to build the new economic value inherent in emerging industries?
4. To what extent will governments, industries and organisations leverage the wealth of capability in Australia's research and development, and academic institutions?
5. How can Australia create a business culture where systems leadership* fosters collaboration across government and industries to help offset transformation risks?
6. What product choices will Australian consumers make that resonate with their rapidly developing climate values. Are they willing to pay a premium for these?
7. What skills and jobs will be required and how do we pursue them to create the lifestyle and society Australians desire?

Disruptive Decarbonisation

All Australians have a role to play in reducing carbon emissions through our choices and actions as we live, work, and play. And we will need to do much more – collectively. But it won't be easy. Therefore, I believe that 'Disruption' is a fitting title for this report as it describes the speed, scale, and impact of interruption Australia requires on climate change.

To achieve our environmental goals will require a very different leadership approach to the decade ahead. More so than ever before in Australia's history.

Ensuring that climate risks and opportunities are appropriately addressed is the important duty of boards of directors and leadership teams. Theirs is the task of long-term stewardship for the companies they oversee. However, to govern climate risks and opportunities effectively, boards and leaders **will need to be equipped with access to better existing and emerging technological insights**, so they can make the rights decisions at the right time.

* Systems leadership is defined as a set of skills and capabilities that any individual or organisation can use to catalyse, enable, and support the process of systems-level change. It combines collaborative leadership, coalition-building, and systems insight, to mobilise innovation and action across a large scale.

Technology

Technology is and will be the single most important contributor to the decarbonisation transformation.

Leaders must embrace and scale the right technologies now if they are to drive sustainable growth for their organisations. And a sustainable future for Australia. The question is:

How to increase Australia's capacity to adapt to a world in accelerated change?

This question has been the subject of my 20 plus years of research on emerging technologies.

There is no doubt that in the coming decade, Australian society will be critically shaped by our response to that question as we now apply it to climate change. This report provides confidence for our future by:

- **Explaining why** an 'exponential' approach, as opposed to 'linear' to change, is critical to decarbonisation transformation to position Australia on the necessary trajectory to achieve its carbon emissions targets. It further suggests that to meet its 2050 net zero target, Australia must reach the 2030 net zero inflection point at a speed of change never achieved before. And the time to act is now.
- **It demonstrates where** organisations need to mobilise boards and leaders to reach the required state of readiness on their environmental programs. Environmental strategies, policies and programs are only prioritised by one in three Australian organisations. One possible explanation for that low rate might be that 'clarity of benefits and unclear ROI' was ranked by board directors and chief executives as their number one concern about their environmental strategic readiness and 0 percent of them reported any increases in profit this year from there environmental programs.
- **It shows how** to create environmental strategies through an evolutionary maturity model that considers Compliance, Risk, and Strategy. Mature environmental programs focus on two key and interrelated dimensions. Firstly, the integration of environmental considerations into strategic decision making and operational execution. Secondly, through a robust environmental goal-setting process that considers stakeholder sentiment, business opportunity and corporate capability.
- **It provides guidance** on how boards and leaders must respond with urgency by adopting and scaling mature and emerging technologies at an unprecedented pace. This is critical as environmental programs are data-intensive, requiring lots of intelligent technology to gather, analyse and leverage the insights from operations, their value chains, and interconnected ecosystems.

There are important insights to discover, compelling comment and content to consider, and strategies to assist Australia's businesses and industries increase their capacity to adapt to a world in accelerated climate change.

1. Introduction

"We are on a highway to climate hell with our foot still on the accelerator"

Antonio Guterres, UN Secretary-General

The Paris Agreement came into force in 2016. Signing that agreement was a major step towards Australia's international efforts to address climate change.

The 2022 United Nations Climate Conference of the Parties (COP27) took place against a backdrop of extreme weather events worldwide, an energy crisis propelled by the war in Ukraine, and scientific data reiterating the fact that the world is not doing enough to tackle carbon emissions and protect the future of our planet.

In November as world leaders, as well as civil society, businesses and concerned citizens, convened in Sharm El Sheikh to accelerate action towards the goals of the Paris Agreement and the UN Framework Convention on Climate Change, UN Secretary-General António Guterres set the scene for COP27.

The three key COP27 objectives to commit to were:

1. **Mitigation:** how are countries reducing their emissions?
2. **Adaptation:** how are countries going to adapt and help others do the same?
3. **Climate finance:** Financial support for vulnerable, developing nations.

On the 16th of June 2022, Prime Minister, Anthony Albanese and the Minister for Climate Change and Energy, Chris Bowen conveyed Australia's updated Nationally Determined Contribution (NDC) under the Paris Agreement to the Executive Secretary of the United Nations Framework Convention on Climate Change (UNFCCC)[1].

It formalised Australia's pledge to reduce greenhouse gas emissions by 43 percent below 2005 levels by 2030, to put Australia on track to achieve net zero emissions by 2050. On 8 September 2022, these targets were enshrined in legislation by Australia's Parliament, providing the certainty industry and investors had been seeking.

The Albanese Government's bold *'Powering Australia Plan'* aims for Australia to become a clean energy superpower and includes:

- Creating more than 604,000 jobs (five out of six in regional areas).
- A $20 billion investment to accelerate the decarbonisation of the electricity grid.
- An additional $300 million to deliver community batteries and solar banks.
- Up to $3 billion investment from the new National Reconstruction Fund to support renewables manufacturing and low emissions technologies.
- Powering the Regions Fund to support the development of new clean energy industries and the decarbonisation priorities of existing industry.
- $100 million to train 10,000 new energy apprentices in the jobs of the future.
- $10 million for a new energy skills program to provide additional training pathways.
- Introducing declining emission baselines for Australia's major emitters, under the existing Safeguard Mechanism.

- Australia's first National Electric Vehicle Strategy, to reduce emissions and accelerate the uptake of electric vehicles (EVs).
- Double existing investment in EV charging and establishing hydrogen refuelling infrastructure ($500 million).
- Applying new standardised and internationally aligned reporting requirements for climate risks and opportunities for large businesses.
- Committing to reduce the emissions of Commonwealth Government agencies to net zero by 2030.
- Restoring the role of the Climate Change Authority, while keeping decision-making and accountability with Government and introducing new annual Parliamentary reporting by the Minister.

Beyond Australia's domestic climate change and energy crisis, the Australian government inherited an unfolding, geopolitically 'fuelled' global energy crisis. This means Australia will need to be at the global macroeconomic table to influence change. Australia's bid to host the next COP, and its offer for Pacific partner countries to co-host, is an important signal about its role both globally and regionally on climate change.

While Australia took a bold range of initiatives to COP27, there remains much to be done and time has run out. This report aims to deconstruct COP27 objectives to an organisational and leader level. It will power them through new insights, augmented with world leading research, on how they can accelerate their organisations environmental programs.

That deconstruction delves into:

- **The recapitalisation risks** associated with the timing of technological investments necessary to become a lower emissions economy given Australia's 'short runway' to 2030.
- **The strategic, investment and operational decisioning making** needed to evolve and mature from climate change action being Compliance Driven to Risk Driven to Strategy Driven. And the need to redress the significant technological under investment Australian organisations now face on climate change.
- **Why the role of the CIO** is critical to orchestrate and choreograph the necessary technological collaboration between government and industry. This will be done through systems leadership and environmental enablement across all functional lines within their organisations.
- **Providing a blueprint** on how to cultivate the skills and leverage existing and emerging technologies, to propel their organisations onto the necessary trajectory to hit 2030 targets and beyond.

2. Recapitalising Australia's Economy

"Transforming Australia to a low carbon economy is inevitable and complex. If we get it right, the opportunities are immense."

Philip Chronican, Chair, National Australia Bank (NAB)

This *Disruptive Decarbonisation* report begins with detailing the significance and magnitude of the investment required to be reallocated and created to decarbonise Australia's economy – whether this investment is either garnished from existing budgets or created. It then identifies which economic systems, and their associated industries require transforming from a carbon intensive to a low emissions state and covers how board directors and chief executives are financially planning their carbon reduction targets. It then shows what appetite they have for paying for future carbon costs and their predictions for when they expect to be forced to pay for it.

2.1. The 4th Industrial Revolution

New research commissioned by NAB and published in July 2022 reveals the economic opportunity, investment, and systems that must be transformed for Australia to transition to a net-zero economy by 2050[2]. The report titled, *'All Systems Go'* by Deloitte Access Economics, estimates that over the next 30 years, Australia is expected to invest around $20 trillion of capital in the economy (net present value).

This is the total level of expected investment in a net zero economy by 2050. It includes investment directly related to the transition such as renewables and investment not directly related, such as education and health. The research suggests that this $20 trillion will need to be invested differently, and in a structurally different way, to compete in a low emissions world.

As with prior industrial revolutions, governments and industries were required to either generate new capital flows or reallocate existing capital to transition to the new economies that industrial revolutions created. The same will be true for Australia in the 4th Industrial Revolution.

> What makes the 4th Industrial Revolution different from others, is the fusion of our digital, physical, biological, and environmental systems requiring an acceleration of capital investments, their magnitude, and the timing of the returns.

For example, the 3rd Industrial Revolution which broadly began in the 1960s, unfolded over 60+ years. But the 'Great Reallocation' of capital required for Australia to decarbonise our economy for this 4th Industrial Revolution, will have to occur in less than half that time - 30 years.

Of the total estimated $20 trillion of capital expected to be invested in the Australian economy by 2050, around $4 trillion is estimated to reflect the capital flows and investments to decarbonise the four core economic systems in the economy: energy, logistics, production, and agriculture.

These systems are not independent but rather they are interconnected and it is how these relationships change that becomes the foundation for the creation of new economic value. The convergence of these systems can be observed is in all prior Industrial Revolutions. Today, the same core economic systems are just characterised by different technologies, scientific advancements, timeframes, and of course proportionate investment levels.

2.2. Systems Transformation

Australia's four core economic systems Deloitte Access Economics describes as requiring transformation to a low carbon emissions economy are energy, mobility, materials manufacturing, and food and land use systems. See **Exhibit 1**.

These core economic systems are supported by a range of enabling services from associated industries including finance, insurance, communications, business, real estate, recreational and other services. These industries also require transformation to a low carbon emissions future.

Exhibit 1: Core economic systems requiring transformation to low emissions

The energy system is critical to how everything in the economy functions. The system's key industries relate to energy generation. But the system also includes upstream industries such as resource exploration and extraction, related manufacturing (e.g., solar panels) and construction (e.g., new generation assets), as well as downstream industries such as electricity distribution.

The mobility system encompasses all transport functions and the inputs required (i.e., fuels and car manufacturing). The transformation of the mobility system will require the near-complete replacement of the vehicle fleet, with the systems approach factoring in the emissions of inputs and processes required to achieve it.

The materials manufacturing system provides vital inputs to all aspects of the economy. The system includes hard-to-abate operations, which currently have few viable low-carbon alternatives. Cement, steel, aluminium, and other hard-to-abate heavy industries could see much wider use of green hydrogen and electrification, with on-site carbon-capture technology playing an important role.

The food and land use system feeds the world. The system contributes significantly to emissions through livestock husbandry, crop burning, and deforestation. The transformation of the food and land use system is expected to see the system go from being a significant contributor to emissions, to a net carbon sink.

Enabling Services are a core input across the transformation of all four core economic systems.

Source: Deloitte Access Economics *'All Systems Go'*

To achieve Australia's emissions targets, Deloitte Access Economics estimates that:

- **By 2030** – $70 billion of capital will need to be reallocated from emissions intensive activities into low emissions activity.
- **By 2050** – $420 billion in new capital investment will be needed to achieve Australia's emissions target. Of which, $400 billion will be required across the four core economic systems and enabling services described in **Exhibit 1**.

Exhibit 2 details the estimated allocation of this capital across the core economic systems and their enabling services, together with today's estimates of their share of emissions and gross value additions to the economy and share of employment.

This information highlights the significant reallocation of capital ($25 billion) from the Energy System by 2030 from carbon intensive to low carbon emissions assets. And the $100 billion of new investment required to transform the Energy System to net zero carbon emissions by 2050.

The mobility, raw materials manufacturing, food and land use systems require much more new capital investment ($100 billion collectively). However, they importantly drive much higher employment and gross value add than the energy system.

To fuel the transformation of the four core economic systems, $200 billion of new capital is the estimate required for transformations across financial, insurance, business, real estate, communication, recreational and other enabling services out to 2050.

Exhibit 2: Total reallocation and new investment in core economic systems and enabling services to meet Australia's targets (net present value) ($billion)(%)

System Contribution and Capital Investment	Energy	Mobility	Raw Materials Manufacturing	Food and Land Use
2022 Share of Emissions	61%	14%	16%	16%
2022 Gross Value Add	23%	17%	31%	13%
2022 Share of Employment	4%	9%	22%	15%
2030 Estimated Capital Reallocation (from emissions intensive to low emissions assets)	$25b	$3.5b	$1.5b	$0.1b
2050 Estimated Capital Allocation (increase in low emissions capital assets)	$100b	$30b	$50b	$20b
2050 – Estimated New Capital for Enabling Services	$200b			

Source: Deloitte Access Economics *'All Systems Go'.* Notes: percentages do not add up to 100 due to overlap of system element. *Enabling services are financial, insurance, business, real estate, communication, recreational and other services out to 2050.*

For the world to reach net zero by 2050, it is predicted that capital spending on equipment and infrastructure with relatively low emissions intensity, would average US$6.5 trillion each year. That is more than 71 percent of the estimated US$9.2 trillion in annual capital spending during that time[3].

In the 2022 Federal Australian Budget, the Government announced that it will roll out its *Powering Australia Plan* to establish a $20 billion fund for energy transmission and invest more than $800 million in a package of measures. These include lower taxes on EVs, building a national EV charging network, and providing solar battery storage for up to 100,000 homes.

2.3. Paying for GHG Emissions

A divided view. While many board directors and chief executives reported having set ambitious carbon emissions reduction targets, many also have either no current target but anticipate setting it in the next 12 months (46%), or have neither set or plan to set a target (40%)(see **Exhibit 3**). Those that have set ambitious carbon reduction targets will require investing in carbon offsetting programs that include targets for:

- **Carbon neutral (50%)**
- **Climate neutral (20%)**
- **Net zero (16%)**
- **Climate positive/negative (4%)**

Whilst the setting of these targets is encouraging, the measurements to support, monitor and report them are far from being standardised, regulated, and supervised. This issue is discussed in detail in **Section 4.5**.

Exhibit 3: Organisations greenhouse gas emissions reduction target

Which of the following greenhouse gas (GHG) emissions reduction targets have been set by your organisation?
(select multiple responses)(%)

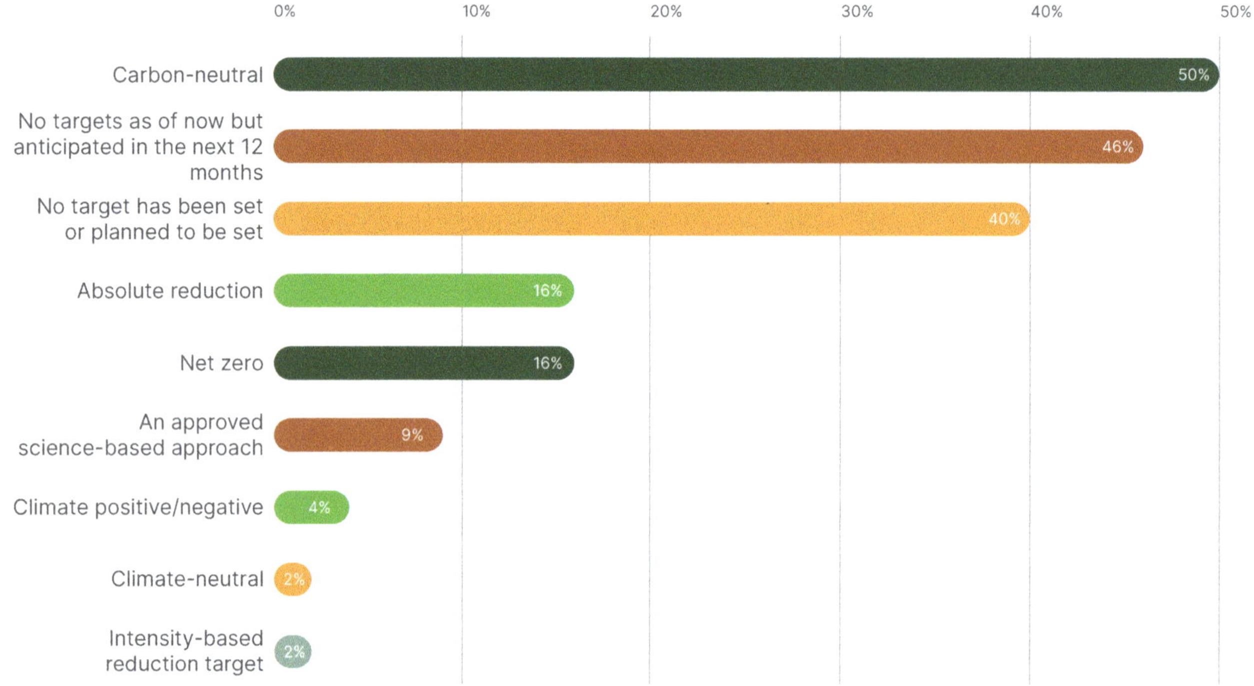

Source: *Australia 2030 Research©* Rocky Scopelliti (n=178)

When it comes to nonvoluntary payment for carbon emissions, 71 percent of board directors and chief executives report that their organisations have not planned for the possibility of future costs associated with carbon emissions by way of some form of carbon taxation or pricing scheme. Other than through voluntary offsets, 65 percent of board directors and chief executives reported that within the next five years, they anticipate being forced to pay for carbon with one in four anticipate expecting to pay within three years (see **Exhibit 4**).

Exhibit 4: Planning for future costs and the timeframe to pay for carbon

Other than through voluntary offsets and investments, how many years from now do you anticipate being forced to pay for carbon?
(select one response)(%)

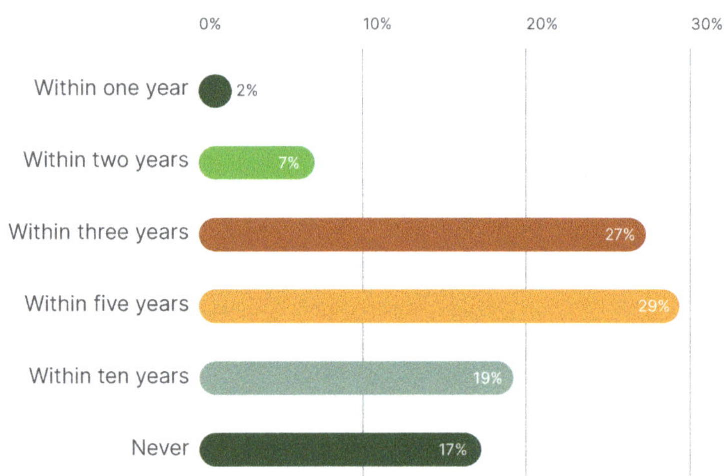

Source: *Australia 2030 Research©* Rocky Scopelliti (n=181)

Does your organisation's financial planning include the possibility of the future costs associated with carbon emission by way of some form of carbon taxation/pricing scheme?
(select one response)(%)

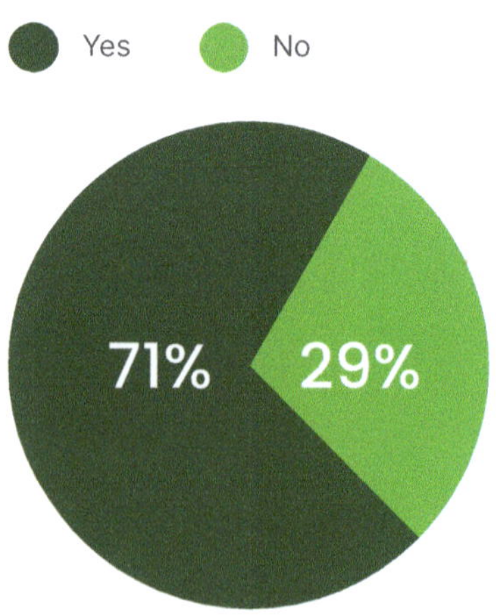

2.4. Section Summary/Key Takeaways

- As with the 1st, 2nd and 3rd, Industrial Revolutions, the 4th will also require reallocation of capital and new capital investments to make the necessary structural adjustments to Australia's economy.
- What makes this 4th Industrial Revolution different is the accelerated timing of the capital investments (reallocated and new), and their magnitude fusing our digital, physical, biological, and environmental systems.
- Systems leadership across all actors in the economy and how they sequence their interdependencies will determine the extent to which Australia optimises this unique opportunity to respond to climate change.
- Systems leadership will be critical to orchestrate the intrinsic link between the four core economic systems, the enabling services, and the necessary transformations to create new economic value. Which in turn is dependent on those transformations occurring exponentially to generate growth.
- The size of the prize of a systems leadership-driven, economic gain for net-zero transformation, is reported to add $890 billion to Australia's GDP by 2070.
- Transforming Australia's energy system will require the highest level of capital reallocation before 2030 and new capital to 2050. Transformation to low emissions assets is also required for the other three core economic systems being mobility, raw material manufacturing, and food and land use. These will also require both a reallocation of capital to 2030 and new capital to 2050.
- $200 billion of new capital investment will be required to transform the four core economic systems, industries and their associated enabling services of financial, insurance, business, real estate, communication, recreational and other services, out to 2050.
- The consequences of an uncoordinated transition, however, will prove more disruptive and expensive. Should the 2050 net zero target not be achieved, it will reportedly attract a $390 billion net present value cost to Australia's economy by 2070.
- In the 2022 Australian Federal Budget announcement, the Government announced it will roll out its *Powering Australia Plan* and establish a $20 billion fund for energy transmission, invest more than $800 million in a package of measures including lower taxes on EVs, building a national EV charging network, and providing solar battery storage for up to 100,000 homes.
- We found that Australian board directors and chief executives seem to be quite divided on setting carbon emissions reduction targets. Many reported that their organisations have either no current target but anticipate setting it in the next 12 months or have not set or plan to set a target.
- When it comes to nonvoluntary payment for carbon emissions, most board directors' and chief executives' organisations reported that they have not planned for the possibility of future costs associated with carbon emissions by way of some form of carbon taxation or pricing. They do expect to be forced to pay for carbon within the next five years.

3. Disruptive Decarbonisation

"Our intuition about the future is linear. But the reality of information technology is exponential, and that makes a profound difference. If I take 30 steps linearly, I get to 30. If I take 30 steps exponentially, I get to a billion."

Ray Kurzweil - world-leading author, computer scientist, inventor, futurist, and co-founder of the Singularity University.

Section 2 of the *Australia 2030 Research©* highlighted the significance of capital investment, the industries impacted, the interdependencies for the decarbonisation transformation ahead and the expectations of their leaders regarding when they anticipate being forced to pay for carbon. That context helps us now examine when those transformations will need to be successfully executed to meet Australia's new environmental goals.

3.1. Exponential Lag Risk

The key issues facing Australia are that accelerating the decarbonisation of its fossil-fuel based prosperity is colliding with a global energy crisis. And the rate of change will continue to accelerate exponentially.

The exponential effect is exemplified by Moore's law[4]. Around the beginning of the 3rd Industrial Revolution, more than 60 years ago, Moore observed that the number of transistors on a single die (a measure of computing power) available at a fixed price, doubled every 18 months or so.

Kurzweil, quoted above, with his 30-year track record of accurate predictions, made four key observations about the rate of change. He described it as *'The Law of Accelerated Change'*[5]. He observed:

1. **The Multiplication Effect.** Just like Moore's law theory of price/performance of computation, this also applies to information technology.
2. **Information Enablement.** Once an industry, product or technology becomes information enabled, it exhibits its own price/performance doubling effect.
3. **Acceleration.** Once this doubling effect begins, it doesn't stop.
4. **Exponential Technologies.** Information technologies including AI, robotics, biotech, and 3D printing, are key enablers of this multiplication and acceleration effect.

> **Information enablement through these accelerating emerging technologies enable a business model superior economic performance whereby the marginal cost of both supply and demand may reduce to virtually zero transforming existing and creating new industries.**

Disruption inflection point: Exponential models deliver the kind of scale that the traditional, linear based models of the past industrial revolutions simply cannot achieve. This is where a traditional value creation system decouples from its relationship to growth, supply of scarce resources, and time. That point is also where the economic inflection points and disruption occur, through the 'law of accelerated returns.' (See number 1 **Exhibit 5**).

This effect is what I refer to as the *'new economic physics'* where our traditional notions of value creation through systems of energy, production, logistics, and agriculture's scalability are decoupled through technological advancement by dematerialisation, disaggregation, and disintermediation throughout entire supply chains.

For many organisations, adapting to the world in a time of accelerated, exponential change, requires an investment in significant and complex capability-building phases. This results in significant transition and transformation risks that often create a disconnect with their technology investment strategies – an 'exponential lag'. (See number 2 **Exhibit 5**).

Timing: To reach the emissions targets by leveraging existing exponential capabilities and investing in new and emerging technologies of the 4th Industrial Revolution, Australian governments, industries, and organisations need to invest now if they are to avoid exponential lag risk to their benefits from decarbonisation (See number 3 **Exhibit 5**).

This means technology investment planning processes need to be reengineered quickly. And done so on the rate at which the organisation can transition and transform to meet the environmental targets. Each organisation's scaling capacity must become a key factor in those processes. They must be able to invest in the relevant technologies more quickly and at the right times in their lifecycle.

> The critical question for board directors and chief executives therefore is: 'Are we accelerating our climate change capability at least as fast as the exponential curve is growing?' If not, they are unlikely to catch up, resulting in the aforementioned 'exponential lag' risk. (See number 2 Exhibit 5).

Exhibit 5: The new economic physics — exponential versus linear transformation relationships, risks, lag, and investment cycle

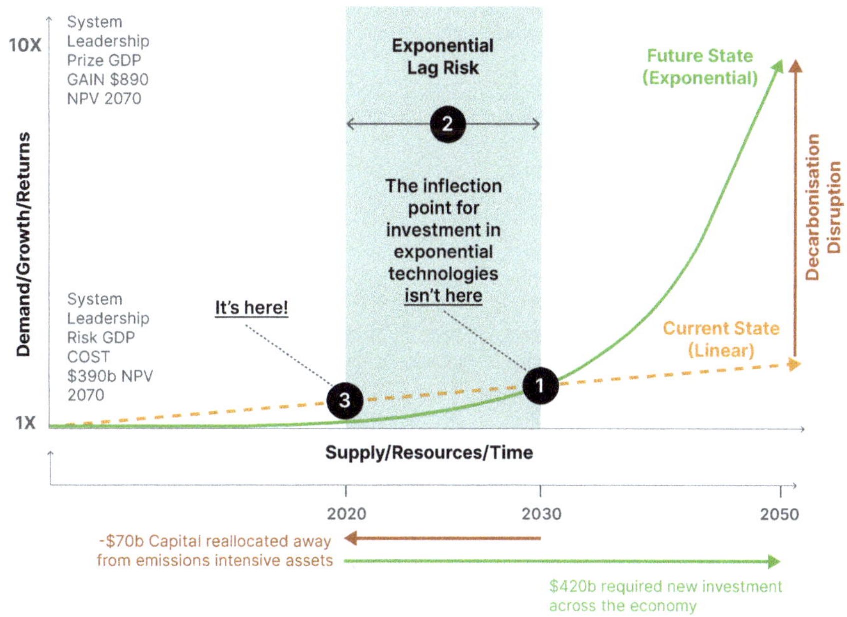

Source: *Australia 2030 Research*© Rocky Scopelliti / Deloitte Access Economics

> Net Zero awaits Australia at the end of the exponential curve, not the linear curve.

For Australian governments, industries and organisations, this rate of change leaves no room for a 'linear' based approach to their climate change transformation. Particularly before 2030. To hit the 2050 target, Australian governments, industries, and organisations will need to hit the 2030 targets on time, and in 2030 they will have to be improving transition performance at least as fast as the exponential curve.

We need to be investing in change now to achieve that outcome.

3.2. Exponential Technologies

While many of the 4th Industrial Revolution's emerging technologies drive exponential performance in of their own right, they become disruptive when combined to create a multiplication effect. In **Exhibit 6** we show a few examples of how some of the information enabled emerging exponential technologies, their characteristics, and their capacity to stimulate the multiplication effect, could impact the achievement of a net-zero future state across industries associated with the four core economic systems and enabling services.

It is the multiplicative impact of these technologies that will transform these industries. The 'enabling technologies' that will be both disruptive and transformative to industries associated with core economic systems and enabling services.

Exhibit 6: Clusters of emerging technologies and their characteristics

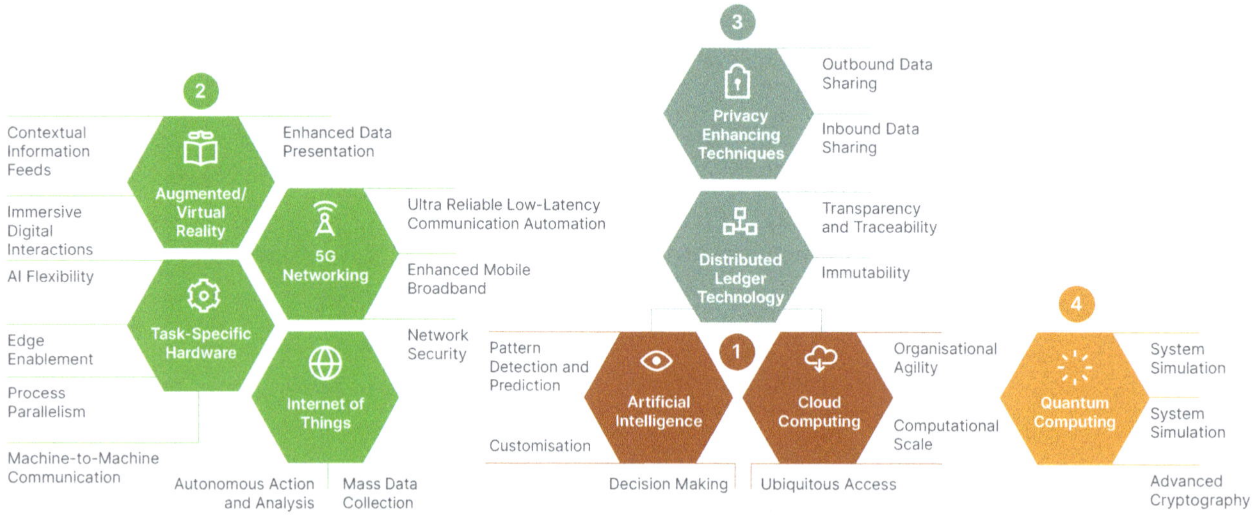

1.

Artificial Intelligence (AI) + Cloud sit at the core of each cluster and are key to unlocking other technologies. AI analyses the data other technologies generate, store and transmit, while cloud is the repository and gateway to accessing technologies-as-a-service.

2.

Internet-of-Things (IoT) devices, connected via 5th Generation (5G) mobile networks, could more directly link physical data to processes. Task specific hardware gives these devices the ability to make preventative decisions at the edge and improve the efficiency of cloud analysis. Augmented Reality/Virtual Reality (AR/VR) create data rich channels for firms to engage with stakeholders.

3.

Distributed Ledger Technology (DLT) will enable more direct, atomic transfers of value and information between parties, in areas where disintemediation is beneficial.

4.

Quantum Computing power, accessed via large cloud providers, will extend the capabilities of AI and unlock specific, powerful new analysis techniques.

Source: WEF: *Australia 2030 Research*© Rocky Scopelliti

There are many other examples of technological and scientific developments associated with the 4th Industrial Revolution that will have specific applicability to the transformation of core economic systems. These are highlighted against the current and future net zero system state described by Deloitte Access Economics in **Exhibit 7**. Some of these when clustered, will be disruptive such as renewable energy + autonomous vehicles + transport-as-a-service.

Exhibit 7: Example Core System 4th Industrial Revolution technological developments

2020		30 Years	2050	
Core Economic System	**Current System State**	**Transformation Requirement**	**Future Net-Zero System State**	**EXAMPLE 4th Industrial Revolution Technological Developments**
Energy	Fossil fuels dominate the energy sector and electricity mix. Renewable sources growing rapidly	• Phase out fossil fuels • Expand renewable energy capacity • Batteries and storage • Transmission infrastructure • Electrification	The decarbonisation of energy sources, the electricity grid enabled fuel switching across the economy and electrification: within households, within businesses and commercial spaces and in industrial and agricultural production.	• Renewable energy generation, capture and storage • Geoengineering • Material science • Bio refineries • Motion energy harvesting • Microgrids • Light energy harvesting
Mobility	Mobility is the largest consumer of energy in Australia. Progress to date has been slow.	• Switch to green fuels • Batteries • Hydrogen • Investment in public transport and micros-mobility infrastructure	Passenger vehicles are electrified, and charging infrastructure is widespread. Consumers preference public transport and micro-mobility. Freight has been decarbonised through electrified rail and alternative fuels.	• Autonomous vehicles • Internet-of-Things (IoT) • Robotic process automation • Holographic technology • Drones and unmanned aerial vehicles • Satellite/5G Mobile networks
Raw Materials Manufacture	Raw materials manufacturing brings significant value to the economy, but it is not yet decoupled from emissions intensity. Fossil fuels dominate energy inputs.	• New production processes • Carbon capture and storage technologies • Power from renewable sources • Less consumption and circular economy	Renewable energy dominates manufacturing fuel inputs and Australia becomes a key player in the global manufacturing market of green steel. Technology offsets emissions in hard-to-abate sectors.	• 3D and 4D Printing • Additive manufacturing • Robotics and Cobots • Digital twins • Nanotechnology • Material science • Machine co-creativity and augmented design
Food and land use	Climatic conditions greatly impact productivity and profit across the system. There remains an adaption and adoption challenge.	• Avoiding deforestation • Adopt sustainable farming techniques • Improve waste • Shifts in consumer food products	Consumers preference sustainable food products and farms have adopted sustainable farming techniques. The rise of the circular economy means waste level is sustainable. The system becomes a net carbon sink.	• Biotechnologies • Cellular agriculture • Vertical farming • Protected cropping • Sustainability by design • Robotics and automation • Genomics and gene editing • Bio refineries • Circular economies • Geospatial biodiversity mapping • Precision agriculture

Source: Deloitte Access Economics: *Australia 2030 Research*© Rocky Scopelliti

3.3. Section Summary/Key Takeaways

- Traditional systems of production, distribution and scale are becoming decoupled from demand, growth and returns, and supply, resources, and time by information enablement from emerging technologies.
- Decarbonisation disruption occurs at the intersection of the linear and exponential curve.
- The critical question for board directors and chief executives is: 'Are we accelerating our climate change capability at least as fast as the exponential curve is growing?' If not, they are likely to suffer an 'exponential lag' risk.
- Getting to net-zero will require a whole of economy transformation. This will require systems leadership between government and industry.
- Scaling capacity must become a key factor to trigger investment in the relevant technologies more quickly, and at the right times in their lifecycle.
- For Australian governments, industries and organisations, this rate of change leaves no room for a 'linear' based approach to their climate change transformations ahead.
- Critically, to hit the 2050 target, Australian governments, industries, and organisations must both hit the 2030 targets on time and improve their transition performance at least as fast as the exponential curve by 2030. To achieve that we need to invest in change now.
- While many 4th Industrial Revolution emerging technologies drive exponential performance, they are more disruptive when combined to create a multiplicative effect.

4. Mobilising boards and leaders to address Climate Change

"The Industrial Revolution went on for a century, but now businesses must recapitalise the economy to zero carbon in less than 30 years. This matters because it creates a very strategic dilemma: How to pivot toward the new, when what you currently have is generating good returns. Especially, where the costs are upfront, and the dividends are back end."

Pradeep Philip, Deloitte Access Economics

The *Australia 2030 Research*© suggests the required rate of change for Australian governments, industries and organisations leaves no room for a 'linear' based approach to their climate change transformations. Given the significance of capital investment, the industries impacted, and their interdependencies, board directors and chief executives need to make decisions unlike many have been accustomed to.

The *Australia 2030 Research*© then further analyses how ready organisations are for that significant decision making. How mature their environmental strategies are, and the extent to which they are factored into board directors' and chief executives' strategic decisions. Lastly, in this section the *Australia 2030 Research*© explores which stakeholders board directors and chief executives are targeting with their environmental strategies, and the benefits they see their organisations realising in 2022.

4.1. The Climate Governance Initiative

The *Climate Governance Initiative* is driven by a community of Non-Executive-Directors focused on making climate change a boardroom priority. Among other priorities it establishes tools and structures for information exchange, decision-making, follow-up, monitoring and reporting. It enables better collaboration between actors by agreeing clear roles, relationships, and responsibilities.

The *Climate Governance Initiative* mobilises boards of directors around the world to address climate change in their businesses. They achieve this by developing and supporting national associations that equip their members with the skills and knowledge needed to make climate change a boardroom priority. It builds on the World Economic Forum's *Principles for Effective Climate Governance* published in 2018. Since then, the initiative has expanded to 56 countries with more than 100,000+ members. [6]

In Australia, the inaugural *Climate Governance Forum* held in August 2022 was hosted by the Australian Institute of Company Directors (AICD). This became the platform where some 1,400 directors and senior business leaders gathered to discuss the effects of climate change, their transition pathways, and investment required, to shift to a net zero economy. It summarised the critical questions raised from the Forum 2022[7] and using the key insights from board directors and chief executives who participated in the *Australia 2030 Research*©, we can see the extent to which the issues and questions raised at that forum concern leaders.

Many Australian organisations and businesses have committed to reaching net zero carbon emissions by 2050. But how achievable is this target? This section investigates that and a range of interrelated questions such as:

- What are leaders biggest concerns about their organisation's readiness?
- What transformations are needed for our economy?
- Do Australia's organisations and businesses have the necessary skills?
- What is the role of the board in setting targets and overseeing progress?
- What are the risks of making, or declining to make, a public commitment?
- And how do they get there?

4.2. Organisational Readiness

Even if an organisation develops a comprehensive environmental program, its objective cannot be to simply react to shifting stakeholder expectations or mandates irrespective of their origin. Nor can it be to address environmental issues ad hoc as they emerge and rely on legal compliance. Mature environmental programs must focus on two key interrelated facets:

1. Integrating environmental considerations into strategic decision making and operational execution.
2. A robust environmental goal-setting process that considers stakeholder sentiment, business opportunity and corporate capability.

As organisations move across this maturity spectrum, their environmental programs move from Compliance-Driven, to Risk-Driven, to Strategy-Driven.

The *Australia 2030 Research*© found that board directors and chief executives ranked 'Climate Change' as their number one issue (35 percent) expected to affect Australia in the next 10 years. This was followed by 'Skills/ Reskilling and the Future of Work' (21 percent).

These concerns highlight the significance of the two interdependent factors for the growth of economy ahead (see **Exhibit 8**).

The urgency and significance needed to prioritise environmental strategy, policies and programs for board directors and chief executives is also highlighted by the finding that 35 percent reported their organisations are yet to establish 'Environmental Strategies, Policies, and Programs' (see **Exhibit 8**). Their concerns about their organisation's environmental strategic readiness could explain this (see **Exhibit 8**).

1. Clarity of benefits/unclear ROI (33 percent).
2. Lack of Skills/Reskilling (23 percent).
3. Under investment in the necessary enabling technologies (19 percent).

Organisational change management' was only reported by 8 percent as a concern, and I suspect that the absence of comprehensive transformational programs and the timing of there execution maybe understating that concern.

Exhibit 8: Issues affecting Australia: Establishing an environmental program and concerns for environmental readiness

In your opinion, which of the following issues will affect Australia over the next 10 years? (rank)(%)

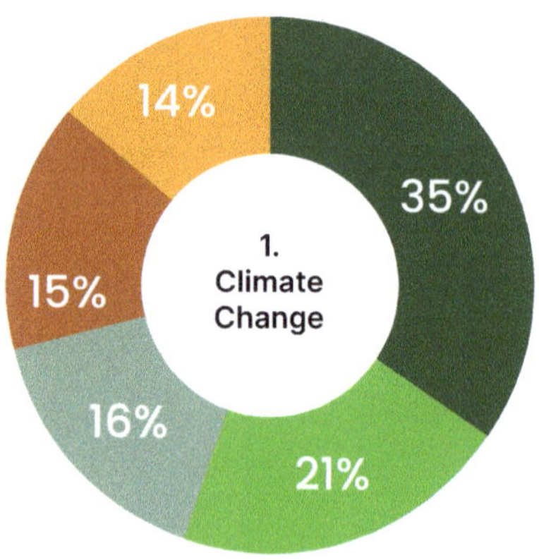

- Climate change and the environment **35%**
- Skills/Reskilling/Future-of-work **21%**
- Geo-Political instability **16%**
- Low economic growth **15%**
- Technological and scientific advancements **14%**

Does your organisation have an environmental strategy, policies and programs?(%)

- Yes **65%**
- No **35%**

What are your top concerns about your organisations environmental strategy readiness over the coming decade? (rank)(%)

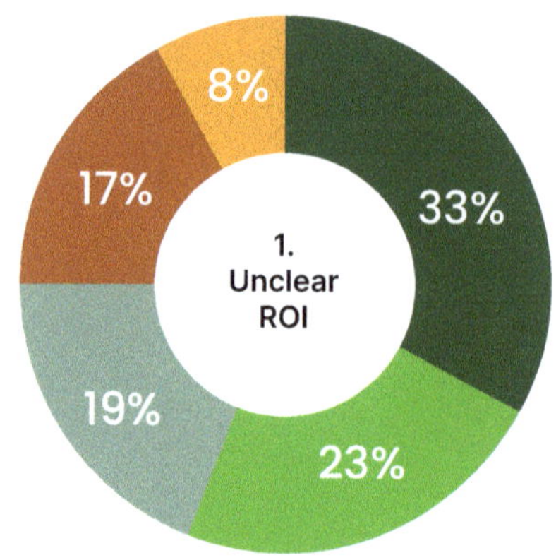

- Clarity of benefits/unclear return on investment (ROI) **33%**
- Lack of skills/reskilling **23%**
- Under investment in the necessary enabling technologies **19%**
- Effective strategy/policy/program execution capacity **17%**
- Organisational change management **8%**

Source: *Australia 2030 Research*© Rocky Scopelliti (n=180)

4.3. Systems Readiness

As **Exhibit 8** shows, overall, board directors and chief executives ranked 'Climate Change and the Environment' as the number one issue expected to impact Australia over the next 10 years.

But when it comes to what concerns them the most about their organisation's readiness for climate change at a systems level, a clearer picture begins to emerge.

Board directors and chief executives from core economic systems industries are significantly more concerned about the 'clarity of benefits and unclear ROI' than the enabling services industries (46 percent versus 26 percent respectively).

This likely reflects core economic systems industries board directors' and chief executives' concerns about the necessary transformations and priorities required for the estimated reallocation of $70 billion capital, to shift from emissions intensive to lower emissions before 2030.

'Skills/Reskilling/Future of Work' was reported as the next most significant issue enabling services industries (26 percent) relative to only 18 percent from those industries associated with core economic systems.

The third most significant concern for board directors and chief executives was 'Under investment in enabling technologies'. This would relate to their primary concern regarding clarity around ROI. Perhaps suggesting that this is what is holding back investment decisions.

Lastly, only one in five board directors and CEO's from both groups, thought their organisations had effective environmental strategies, policies and programs(see **Exhibit 9**).

Exhibit 9: Top concerns of board directors and chief executives for their organisation's environmental readiness over the next decade

Top Concerns of Board Directors and Chief Executives from industries in Core Systems and Enabling Services regarding their organisational environmental strategic readiness this decade (%)

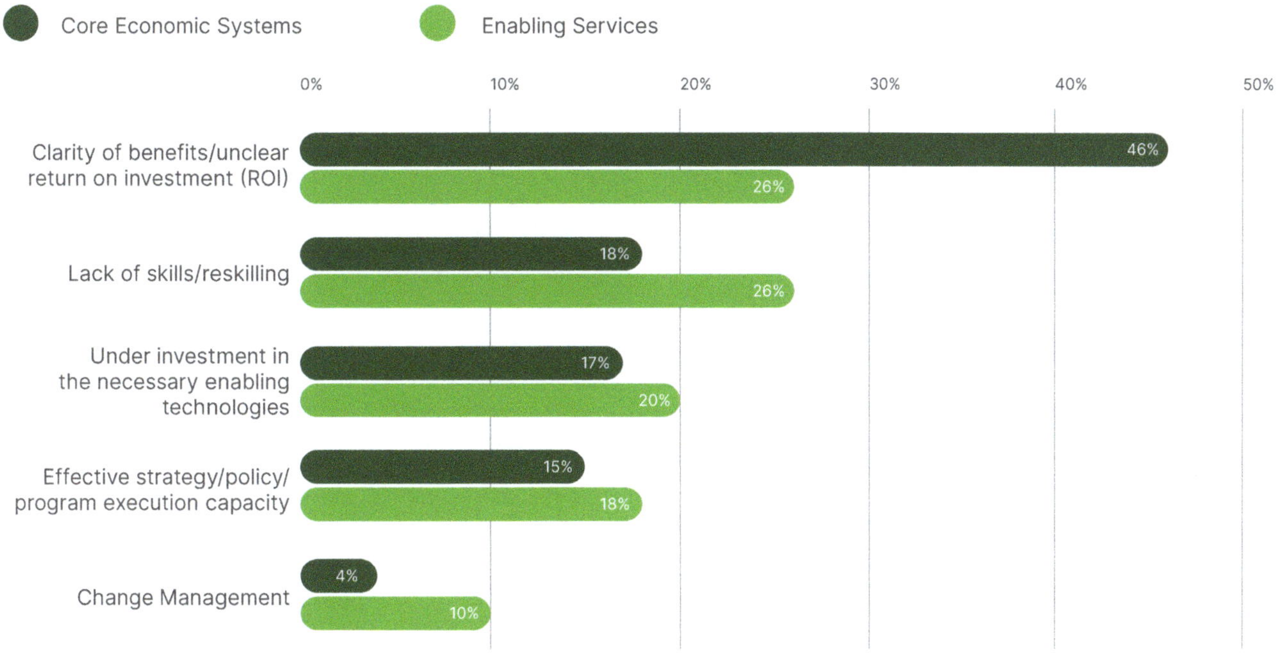

Source: *Australia 2030 Research*© Rocky Scopelliti (n=162)

4.4. Tracking progress and success: climate disclosure and reporting

All entities are under increasing pressure to publicly disclose action on climate change. Discussions and government consultations are underway for a compulsory disclosure framework both globally and in Australia. The questions posed include:

- How might reporting and disclosure obligations drive changes to an organisation's climate actions?
- Is strategy, risk or compliance driving our decision making and goal setting?
- What information are investors and other stakeholders likely to require?
- How will directors address the risks of greenwashing?
- How will boards obtain reasonable assurance of outcomes?

It is important to point out that these questions are not just issues of technology or the will to implement. In many cases, the science is still being developed. For example, in agriculture, there are many significant current global research programs in place to establish agreed models to estimate soil's carbon holding capacity and carbon biomass.

Mandatory climate reporting is expected to be an early focus of the Albanese government. So, board directors and larger company chief executives should be preparing for that. It is likely that any reporting requirements will closely follow the Climate Standard currently being consulted on by the International Sustainability Standards Board (ISSB).

It is likely that the ISSB will be used as the basis for any Australian standard on climate or broader sustainability. It is also expected that the implementation process will start with large, listed companies and heavy carbon emitters as in the case of climate disclosure.

The 2022 Australian Federal Budget earmarked $36.1 million in spending over the next four years (and $6.9 million per year ongoing) in support of government priorities on climate modelling and reporting standards. This includes funding Treasury to restore the capability to model climate risks and opportunities. Treasury and the AASB are reported to receive $6.2 million over four years to develop and introduce specific climate reporting standards for large businesses and financial institutions, in line with international standards.

The *Australia 2030 Research©* revealed that board directors and chief executives are only in the very early stages of 'tracking emissions and energy impact in real-time' or have 'embedded emissions and power monitoring across supply chains' (13 percent).

Leaders reported that most of their organisations (65 percent) have either 'Not Commenced (28 percent) or have a 'baseline established across the organisation' (37 percent)(see **Exhibit 10**).

When it comes to the extent leaders see the integration of environmental considerations into strategic decision-making and operational execution, the *Australia 2030 Research©* revealed that the majority of board directors and chief executives say they are Compliance (32 percent) or Risk (25 percent) Driven. Approximately 43 percent reported that their organisations were 'Strategy-Driven', where environmental matters are integrated into corporate strategy, operations, and scenario planning (see **Exhibit 10**).

Looking at whether organisations have established a robust environmental goal-setting process that considers stakeholder sentiment, business opportunity and corporate capability, board directors and chief executives reported they are Compliance (36 percent) and Risk (25 percent) Driven. Approximately 39 percent reported that their organisations were 'Strategy Driven' where they develop a goal setting process and consider the value and competitive advantage of simply meeting environmental expectations or setting leadership standards for each issue (see **Exhibit 10**).

Exhibit 10: Maturity stages of environmental strategies / Degree of integration into decision making / Robustness of goal setting process

How mature is your organisation's environmental strategies. policies and programs? (select one response) (%)

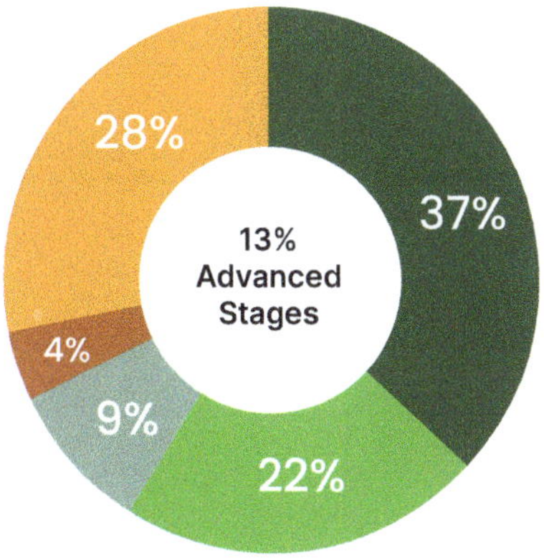

- Stage 1. Baseline established across the organisation **37%**
- Stage 2. Measuring and reporting against baseline **22%**
- Stage 3. Tracking emissions and energy impact in real-time **9%**
- Stage 4. Embedded emissions and power monitoring across supply chain **4%**
- Stage 0. Not yet commenced **28%**

How would you describe the integration of environmental considerations into strategic decision making and operational execution? (select one response)(%)

- Compliance-Driven (e.g. addresses environmental issues piecemeal as they emerge) **32%**
- Risk-Driven (e.g. considers environmental issues in strategic planning discussions as a means of mitigating risk) **25%**
- Strategy-Driven (e.g. intergrates environmental matters into corporate strategy, operations and scenario planning) **43%**

Has your organisation established a robust environmental goal-setting process that considers stakeholder sentiment, business opportunity and corporate capability? (select one response)(%)

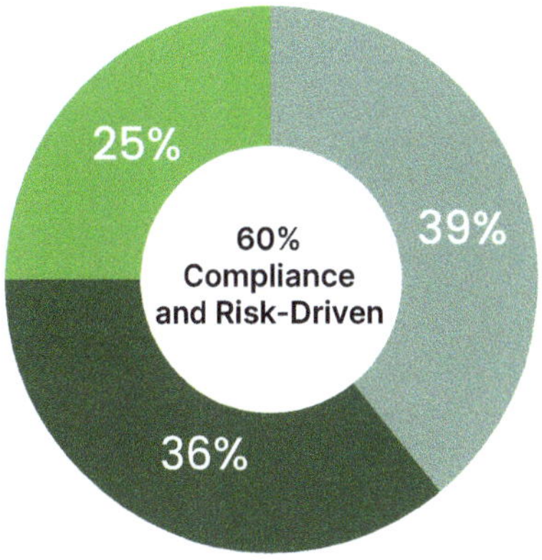

- Compliance-Driven (e.g. indiscriminately and exclusively complies with all environmental regulations and mandates) **36%**
- Risk-Driven (e.g. creates environmental goals based upon risks presented, stakeholder expectations and regulations) **25%**
- Strategy-Driven (e.g. develops a goal setting process and considers the value and competitive advantage of simply meeting environmental expectations or setting leadership standards for each issue) **39%**

Source: *Australia 2030 Research©* Rocky Scopelliti (n=182)

4.5. Systems Environmental Maturity Stage

At a system and industry level overall, up to one in three board directors and chief executives from core economic systems industries and enabling services (refer **Exhibit 11**) reported their organisations are yet to commence their environmental strategies, policies, and programs.

This highlights the significant, and urgent transformational planning effort required to establish a baseline across their organisations. Industries associated with core economic systems are much more advanced in stages 2, 3 and particularly so with stage 4 relative to those industries associated with enabling services (see **Exhibit 11**).

Exhibit 11: Maturity stage of industries from core economic systems and enabling services categories

Maturity Stage of Board Directors and Chief Executives Organisations on Environmental Strategies, Policies and Programs (%)

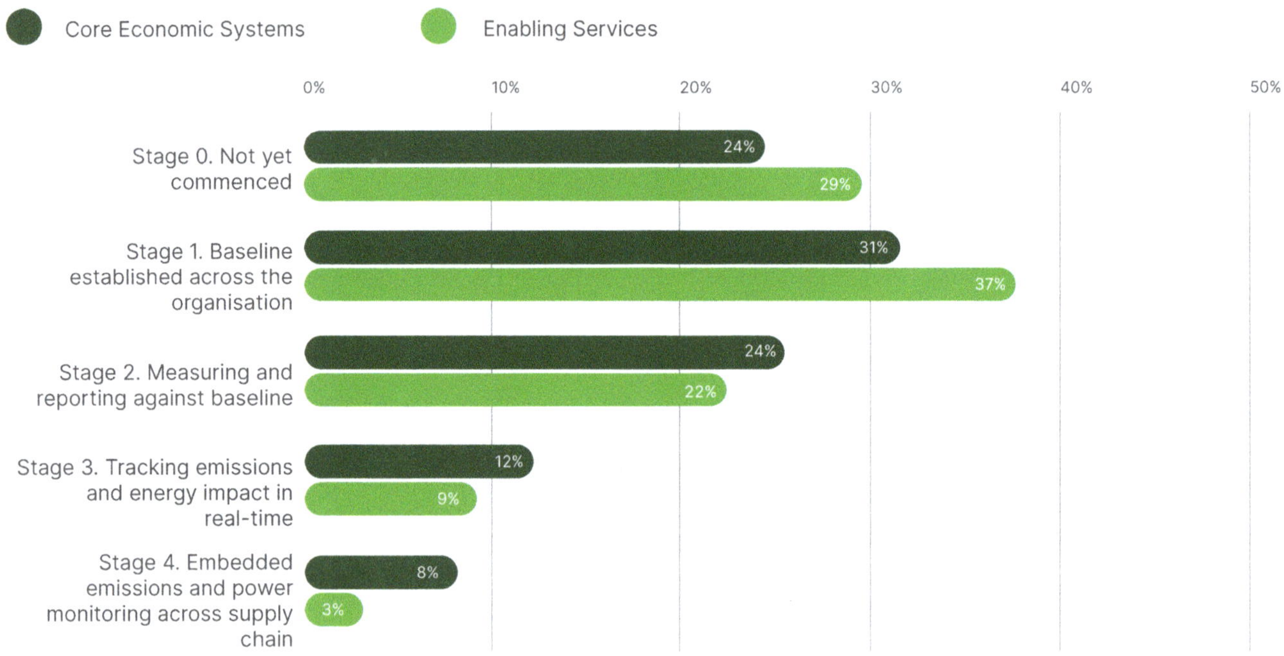

Source: *Australia 2030 Research*© Rocky Scopelliti (n=183)

4.6. Environmental integration within decision making

The extent to which board directors and chief executives integrate environmental considerations into strategic decision making and operational considerations also reveals some interesting insights.

Overall, the *Australia 2030 Research*© found that leaders collectively were mostly 'Strategy-Driven' (89 percent), followed by 'Compliance-Driven' (61 percent). With 50 percent being 'Risk-Driven' (see **Exhibit 12**).

The degree to which environmental considerations are integrated into decision-making is more Compliance and Strategy Driven, than Risk Driven. This may further explain why, as shown in **Exhibit 9**, board directors' and chief executives' main concerns about their organisational environmental readiness is 'clarity of benefits and unclear ROI.'

When it comes to decision making at a systems and industry level, 48 percent of board directors and chief executives collectively from core economic systems industries reported being 'Strategy-Driven'. They integrate environmental considerations into strategic decision-making and operational considerations. In contrast, 27 percent were 'Risk-Driven' and consider environmental issues in strategic planning discussions as a means of mitigating risks. A quarter were 'Compliance-Driven' and addressing environmental issues piecemeal as they emerge.

For those industries classified as enabling services, a similar profile emerges. With 41 percent reporting to be 'Strategy-Driven', 23 percent 'Risk-Driven' and 36 percent 'Compliance-Driven' (see **Exhibit 12**).

Exhibit 12: Degree of decision-making integration of industries from core economic systems and enabling services categories

Degree of integration of environmental considerations into strategic decision making and operational execution (%)

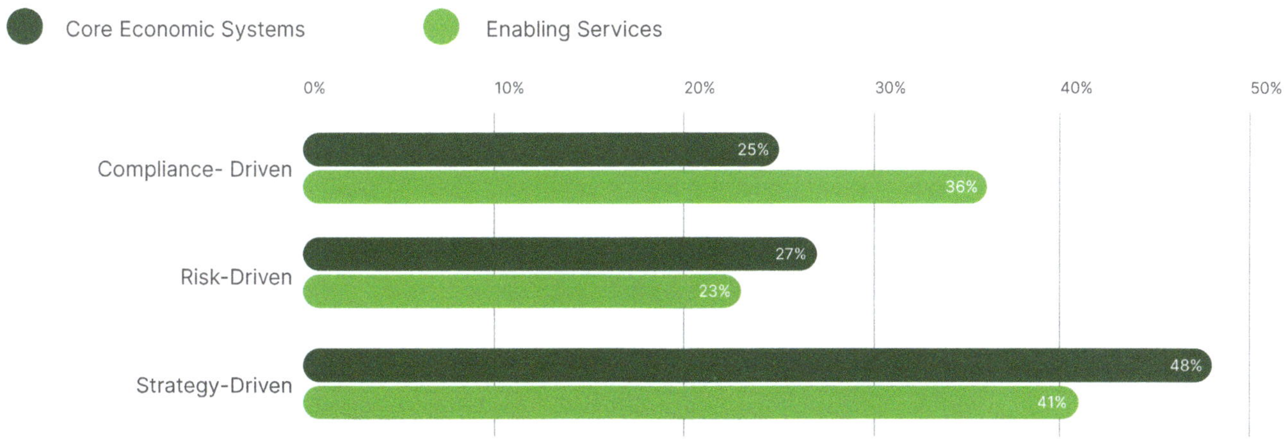

Source: *Australia 2030 Research*© Rocky Scopelliti (n=182)

The extent to which board directors and chief executives see their organisations have established a robust environmental goal-setting process that considers stakeholder sentiment, business opportunity and corporate capability, reveals some interesting insights.

Overall, the *Australia 2030 Research*© found that they were collectively and principally 'Strategy-Driven' (86 percent), followed by 'Compliance-Driven' (68 percent), with 49 percent being Risk-Driven' (see **Exhibit 13**). As with decision-making, discussed earlier, the fact that 'Risk-Driven' was reportedly lower than Compliance and Strategy-Driven, may again explain why 44 percent of boards see 'clarity of benefits and unclear ROI' as the main concern about their organisational environmental readiness.

It also may further explain why 'under-investment in the necessary enabling technologies' was the next major concern for those from the core economic systems industries (24 percent), and the 'lack of skills/reskilling' to be next for those from the enabling services industries (36 percent).

For industries associated with the core economic systems, 45 percent reported to be 'Strategy-Driven', where their organisations have established a robust environmental goal-setting process that considers stakeholder sentiment, business opportunity and corporate capability. In contrast, the 29 percent that reported being 'Compliance-Driven' indiscriminately and exclusively comply with all environmental regulations and mandates. Approximately 26 percent reported to be 'Risk-Driven' and create environmental goals based on the risks presented, stakeholder expectations and regulations.

For the enabling services industries a very different profile is observed. 41 percent report being 'Strategy-Driven', 24 percent 'Risk-Driven', and 39 percent 'Compliance-Driven' (see **Exhibit 13**).

Exhibit 13: Degree of environmental goal setting of industries from core economic systems and enabling services categories

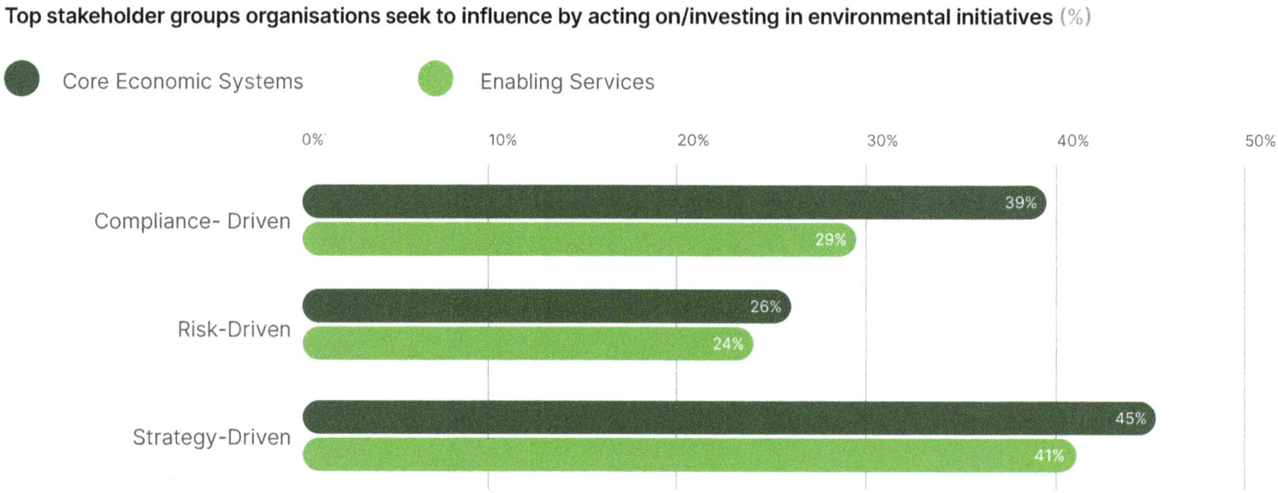

Source: *Australia 2030 Research©* Rocky Scopelliti (n=182)

4.7. Delivering for stakeholders – customers, employees, and investors driving change

Many boards report that stakeholders expect their organisations to take firm action on climate change. Investors, employees, customers, and the broader community all want boards to deliver real outcomes. How can directors and leaders better understand the needs of stakeholders regarding climate change? What role should stakeholders have in a board's decision-making process? What are the risks of not adequately considering stakeholders' interests?

The term 'greenwashing' applies to the actions of a company that gives the impression of acting in an environmentally aware and sustainable fashion while merely 'ticking the boxes' for marketing and reputation purposes. Intense scrutiny and action against greenwashing is escalating worldwide.

Organisations and their leaders are on notice regarding overstatements or misleading claims that could incur potential reputational damage and regulatory risk, as highlighted by the AICD Case Research[8] (see **Exhibit 14**).

Exhibit 14: Santos Case Study

Oil and gas giant Santos was in the headlines as the first company in the world to have a legal case launched against it challenging the accuracy of its net-zero emissions target. The Environmental Defenders Office (EDO) brought the case on behalf of Santos shareholder, the Australasian Centre for Corporate Responsibility (ACCR). The case challenged Santos' claims that natural gas is 'clean fuel' and that it has a credible pathway to net-zero emissions by 2040.

Source: *Australian Institute of Company Directors*

In Australia, the Australian Competition and Consumer Commission (ACCC) is cracking down on greenwashing by Australian companies. It is on the lookout for companies making false claims about environmental action. This follows a global investigation by consumer protection groups that found as many as 40 percent of environmental claims may be fraudulent[9].

In Australia, that figure remains unknown due to the relative absence of transparent climate reporting requirements and government frameworks. However, Gina Cass-Gottlieb, chair of the ACCC, told a House of Representative hearing in October 2022 that 'increasingly large proportions of consumers were making decisions based on products' sustainability credentials'.

Former ACCC Deputy Chair, Delia Rickard further explained that "broad terms like environmentally friendly, green, or sustainable, have limited value and may mislead consumers"[10]. Rickard said that "businesses should provide evidence to back up their claims, whether through reliable scientific reports, transparent supply chain information, or reputable third-party certification." Adding that: 'We will be asking businesses to substantiate their claims'.

Tlou Energy was fined $53,280 in October 2022 in ASIC's first greenwashing penalty and became the first from a major Australian regulator cracking down on this type of misconduct[11]. Others such as Vanguard Investments Australia Ltd have since followed. It raises the questions as to whether 'The fines are sufficiently meaningful enough to correct behaviour? Or small enough to just be a cost of doing business?'

The ***Australia 2030 Research***© looked at the top stakeholder groups that board directors and chief executives reported their organisations seek to influence by acting on and investing in environmental initiatives.

Overall, the ***Australia 2030 Research***© found that board directors and chief executives see 'Customers' (95 percent), 'Employees' (38 percent), and 'Investors' (23 percent) as the top three stakeholder groups their organisations seek to influence (see **Exhibit 15**).

Exhibit 15: The top stakeholder groups the organisation seeks to influence on environmental initiatives by industries from core economic systems and enabling services categories

The top stakeholder groups the organisation seeks to influence by acting on/investing in environmental initiatives (%)

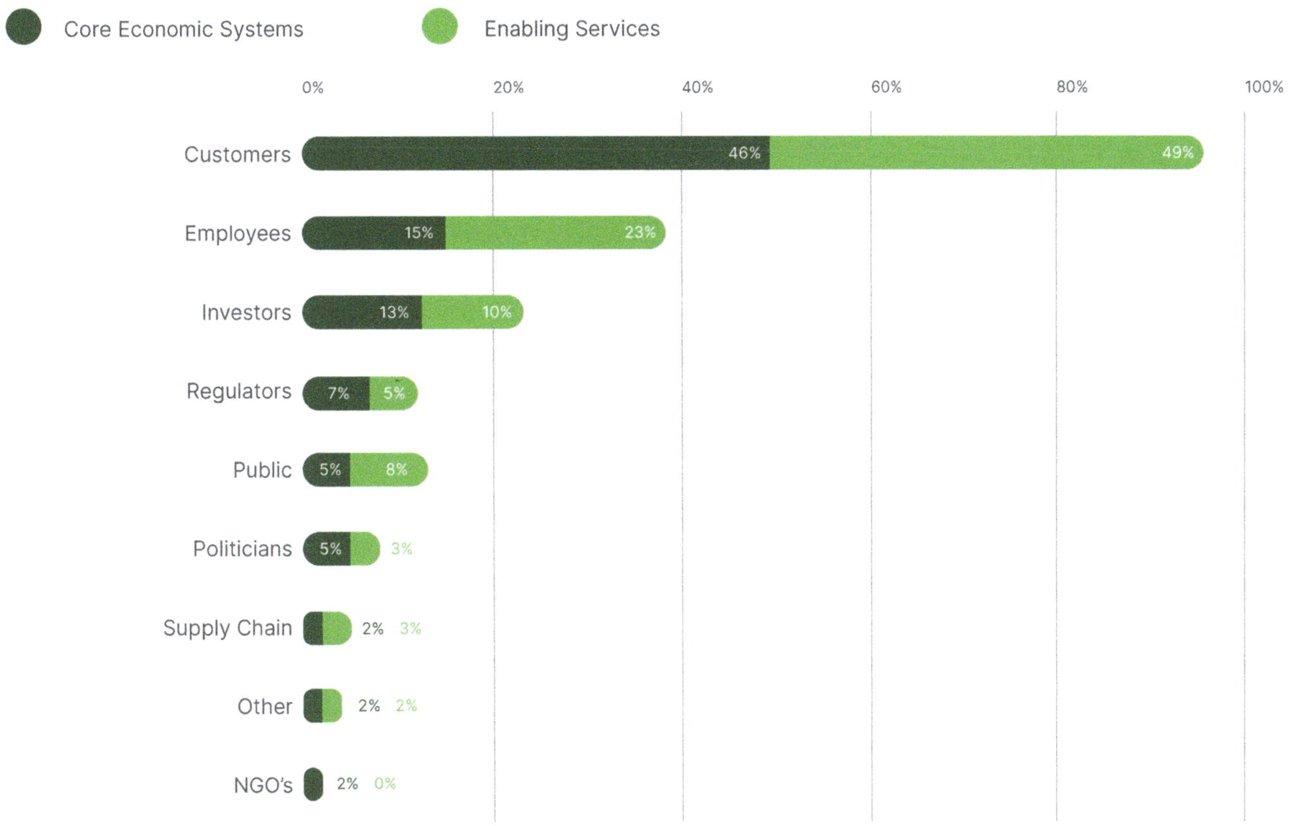

Source: *Australia 2030 Research*© Rocky Scopelliti (n=180)

At a systems and industry level, the *Australia 2030 Research*© found that no material differences between the stakeholder groups by those industries associated with core economic systems or enabling services – other than 'employees'. Enabling services industries rated 'employees' 8 percent more highly than those from core economic systems industries (see **Exhibit 15**)

The research reveals the level of effort required by all Australian industries on supply chains. As previously mentioned, systems leadership and collaboration will be critical to all industry sectors over the coming decade and beyond if they are to benefit from the transition into a low carbon economy. Yet supply chains were rated second last as a stakeholder group to influence on environmental initiatives. This is a major issue as the ACCC expects that 'transparent supply chain information' will be critical for organisations to back up their carbon emissions claims going forward.

This ranking of 'stakeholders' helps to explain the dilemmas faced by board directors and chief executives around 'clarity of benefits and unclear ROI'. And their main concern about their organisation's strategic readiness over the coming decade.

When we look at the question of 'what they believe the top benefits they have already achieved from their organisation's environmental program(s) in 2022', the *Australia 2030 Research*© found that overall, 47 percent believe that beyond Compliance, their organisations have not received any benefits (see **Exhibit 16**). This finding was much greater for enabling services industries than those from core economic systems (29 percent versus 18 percent respectively).

On the other hand, 'Increasing the Company's Brand and Reputation' was the third highest (32 percent) category the participants believe they have already benefited from their organisation's environmental program(s) in 2022. This was across both industries associated with core economic systems and enabling services.

This may further explain the attention that Australian regulators are paying to 'greenwashing' – particularly when only 13 percent reported that their organisations have increased 'Customer Satisfaction/Engagement' as a benefit of their programs in 2022. And that 11 percent reported 'Improved investor engagement and stock market valuation' as the top benefits realised.

Lastly, the focus on 'increasing the company's brand and reputation' is not translating into 'increased revenue'. Only 2 percent of the core economic systems industries and 0 percent from enabling services reported increased revenue. This may further explain the concerns that board directors and chief executives have with growth related strategies when they are highly concerned by 'clarity of benefits and unclear ROI'.

Exhibit 16: The top benefits achieved by organisations in environmental programs by industries from core economic systems and enabling services categories

What are the top benefits that have already been achieved from your organisations environmental program(s) in 2022 (%)

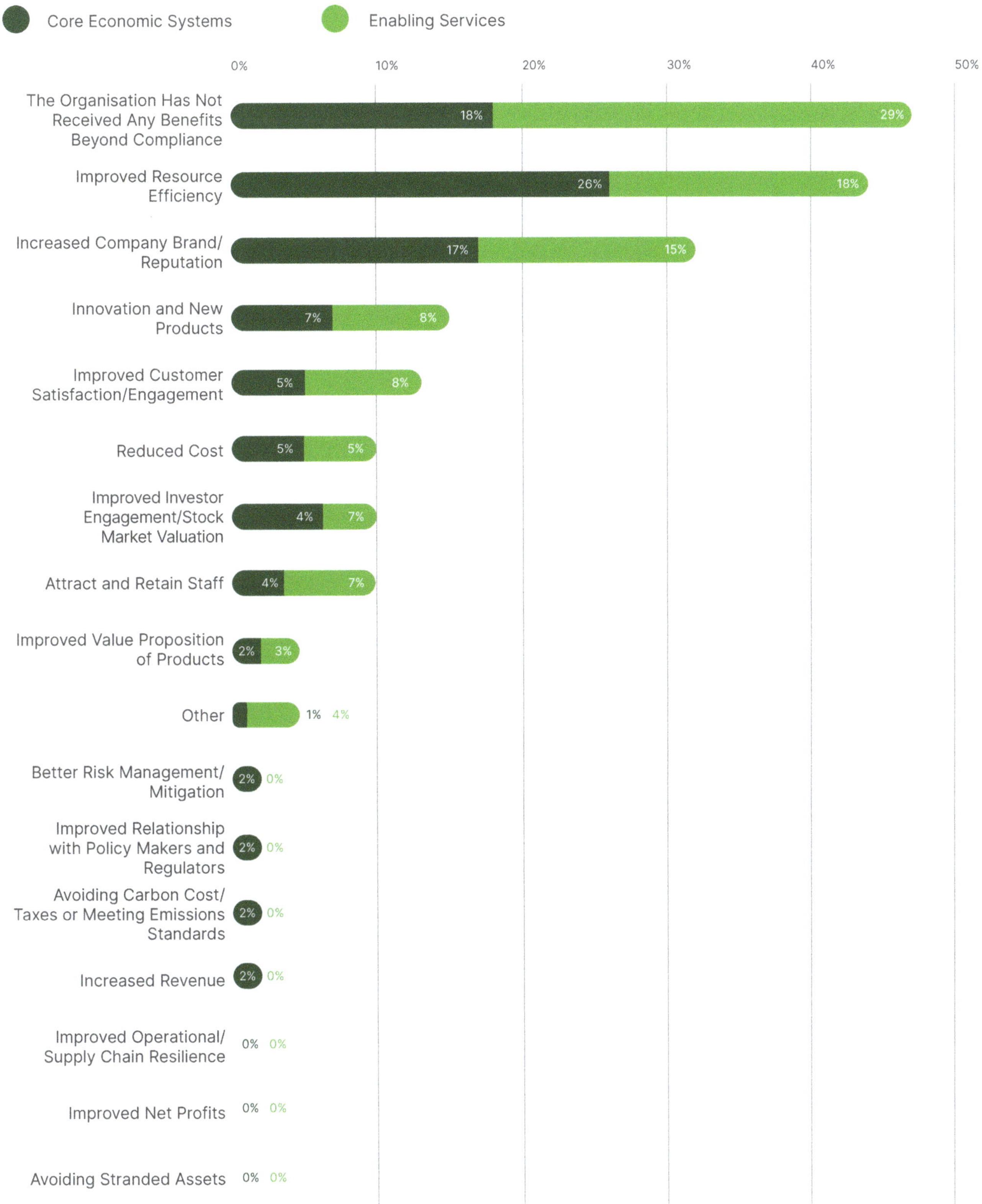

Source: *Australia 2030 Research*© Rocky Scopelliti (n=172)

These insights suggest that Australian organisations need to do much more work around clarifying the benefits to the organisation from their environmental programs. Particularly so when it comes to revenue generation, risk management, avoiding stranded assets, supply chain resilience and net profit. All benefit drivers that board directors and chief executives rated the lowest in terms of the benefits arising in 2022 from their environmental programs.

With new environmental targets delivering inevitable changes to all industries, organisations, workplaces and consumers, greater attention on the opportunities organisations have on 'innovation and new products' and 'improving the value proposition of products' may provide a rationale for growth related investment.

For growth, only 2 percent of organisations from core services reported 'increased revenue' as one of the benefit drivers. Neither industries associated with core economic systems nor enabling services categories, attributed 'improved net profits' as a benefit to their organisations in 2022.

The critical issue, however, is that beyond compliance, most boards and chief executives are yet to see any growth-related benefits from their environmental programs. These findings support the dilemma posed by Pradeep Philip from Deloitte Access Economics in the opening quote of **Section 4** "How to pivot towards the new, when what you currently have is generating good returns, where costs are upfront, and the dividends are 'backend'.

4.8. Section Summary/Key Takeaways

- Board directors and chief executives see climate change as the number one issue expected to affect Australia over the coming decade, followed by skills, reskilling and the future of work. This highlights the significance of these two interrelated factors for the economy in the next ten years.
- For one in three board directors and chief executives reported that their organisations that have no environmental strategy, policies, and programs today. One possible explanation of that low rate might be their number one ranked concern around 'clarity of benefits and unclear ROI' about their environmental strategic readiness.
- Board directors and chief executives reported that their organisations were only in the very early stages of maturity when it comes to environmental, policy and programs. And when it came to 'tracking emissions and energy impact in real-time' or having 'embedded emissions and power monitoring across supply chain' most organisations have either 'Not Commenced' or have a 'Baseline established across the organisation'. This is a major concern as regulators are expecting businesses to provide evidence to back up and substantiate their claims, whether through reliable scientific reports, transparent supply chain information or reputable third-party certification.
- When it comes to the integration of environmental considerations into strategic decisions making and operational executing, the majority of board directors and chief executives see that they are principally Compliance or Risk Driven. Most reported that their organisations were 'Strategy-Driven' where environmental matters are integrated into corporate strategy, operations, and scenario planning. Further, the imminent carbon emissions reporting as part of an organisations financial requirements.
- Board directors and chief executives reported that their organisations are principally Compliance and Risk Driven when it comes to establishing robust environmental goal-setting processes that considers stakeholder sentiment, business opportunity and corporate capability.
- When it comes to the integration of environmental considerations into strategic decision making and operational execution considerations, the *Australia 2030 Research*© found that board directors and chief executives were principally 'Strategy-Driven', followed by 'Compliance-

Driven', and lastly Risk-Driven'. The relatively lower reporting of risk as a driver may explain why boards see 'clarity of benefits and unclear ROI as one of their main concerns about their organisational readiness.

- Regarding the extent to which board directors and chief executives see that their organisations have established a robust environmental goal-setting process that considers stakeholder sentiment, business opportunity and corporate capability, the *Australia 2030 Research*© found that they were principally 'Strategy-Driven', followed by 'Compliance-Driven', and lastly Risk-Driven'. As with decision-making, the fact that Risk-Driven was reported lower than 'Strategy' or 'Compliance-Driven', may explain why boards see 'Clarity of benefits and unclear ROI as the main concern about their organisational environmental readiness.
- Board directors and chief executives reported that the top stakeholder groups that organisations seek to influence by acting on/investing in environmental initiatives, were 'customers', followed by 'employees', and 'investors'. However, the results reveal the level of systems leadership effort required by all Australian industries on supply chains, which were rated second last for stakeholder influence.
- The dilemma faced by board directors and chief executives on why the 'clarity of benefits and unclear ROI' is their main concern about their organisation's environmental strategic readiness over the coming decade, maybe found in the benefits. Beyond compliance, for 2022, board directors and chief executives reported their organisations 'Have not received any benefits'.
- However, 'Increasing the company's brand and reputation' was reported by board directors and chief executives as the third highest benefit category achieved from their organisation's environmental program(s) in 2022. This may further explain the increased attention that Australian regulators are paying to greenwashing – particularly when only 13 percent of board directors and chief executives believe they have increased 'customer satisfaction/engagement' as a benefit of their programs in 2022. Lastly, the focus on 'increasing the company's brand and reputation' was not reported to translate into 'increased revenue'.

5. The CIO as the Enabler and Systems Leader of the Environmental Strategy and Initiatives

"The greatest danger in times of turbulence is not the turbulence – it is to act with yesterday's logic."

Peter Drucker – Management Consultant, Academic, Author

Digital transformation and environmental initiatives are becoming inextricably linked. With technology leaders playing a key role in both. As outlined in the previous section, one of the key requirements for any organisations environmental strategy is the ability to collect, manage and analyse vast amounts of data that will help them ultimately direct, monitor and report its financial and environmental impact.

As shown in **Exhibit 11**, 53 percent of board directors and chief executives reported that their organisations are not measuring carbon emissions today. The *Australia 2030 Research*© suggested that, moving forward, this will require implementing the systems needed to perform those functions. This section will argue that technology leaders have a key role to play as the orchestrator of transformational strategy, while also choreographing the complex dependencies required to execute the organisation's environmental programs and initiatives.

The *Australia 2030 Research*© uncovered the insight that while most Australian organisations' climate goals are focused on the 'customer', from a benefit perspective in 2022, most board directors and chief executives saw that coming from 'increasing the company's brand and reputation' and not from increased revenue or profit.

However, as Australian industries and organisations evolve into the next phase of designing, planning, building, testing, and executing their broader environmental strategies, having transformational leadership will become critical. As such, in this section, we will take a close look at the role of technology leaders as systems leaders and environmental enablers.

5.1. The role of the CIO

Behind 'Climate Change', board directors and chief executives reported 'Skills, Reskilling, and Future-of-Work' as their next greatest concern about the decade ahead. Both these issues are inextricably linked, which inevitably results in consideration of the role of the Chief Information Officer (CIO) and why it needs to evolve.

The CIO emerged as a job title in the 1980s. Back then, this highly technical role would oversee the information technology department's resources and staff. But in the decades that followed, CIOs themselves rewrote their roles. They emerged from the back office to become true strategic business partners, working alongside their C-suite peers. Today, CEOs look to the CIO to drive innovation and revenue-generating ideas throughout the business.

CIOs drive the digital transformation throughout the organisation and wear multiple hats – from business strategist, steward, technologist, catalyst, and change leader. The role of a CIO over the years has evolved from primarily managing technical projects, to influencing major business decisions in an organisation. And it is for these very reasons, I propose that the role of the CIO will evolve again to become the 'Chief Systems and Environmental Officer'.

This new role will be critical in orchestrating climate change transformation across Australian organisations – and more widely across industries and governments over the coming decade.

Although the CIO's responsibilities can vary based on their industry, organisational size, and complexity, many of their general duties include:

- Overseeing the Information Technology (IT) department(s) and setting strategy and direction.
- The enterprise technology, digital, data, engineering, architecture and security strategy and functions.
- Managing IT staff and developing department goals.
- Developing and overseeing the IT budget.
- Planning, deploying, and maintaining IT systems and operations.
- Managing the organisation's software development needs.
- Developing IT policies, procedures and best practices that meet regulatory requirements.
- Directing and overseeing IT projects, digital transitions, and transformation.
- Staying updated on IT trends and emerging technologies.
- Developing and enforcing IT best practices across the organisation.
- Ensuring that IT strategies and processes support company-wide goals.
- Overseeing relationships with vendors, contractors, and service providers.
- Leading the business case to the board of directors and other executives, including the benefits and risks of data, security, and IT-related projects.

Environmental programs are both data and technology intensive and their sources are diverse and highly fragmented. The *Australia 2030 Research©* invited participants to briefly describe the role of the Chief Information/ Technology Officer or equivalent in supporting their organisations environmental program(s). A range of attributes were clustered around the role of the CIO as a systems leader and that as an environmental enabler.

5.2. The CIO as the Systems Leader

When comparing responses from board directors and chief executives to CIOs or equivalents, and participants from all other functions, we see significant opportunities to leverage the CIO leadership role. That is to lead and execute an organisation's environmental strategies and programs.

For example, the research identified that when it comes to the CIO in the role as the systems leader for an organisation's environmental programs, perceptions of the key characteristics and the role itself varied widely between board directors and chief executives, CIOs/Technology leaders and function leaders and a range of systems leadership characteristics as to how the role of the CIO is seen (see **Exhibit 17**).

At an organisational environmental strategy and program level, only one in five describe the role of the CIO, 'as an advisor or part of the environmental strategy team' or 'developing green IT strategy and green IT training'. And only one in three see the role of the CIO in 'supporting the organisations supply chain management' or managing 'cross functional relationships, alignment and communicating the organisations environmental plans' (see **Exhibit 17**).

Exhibit 17: Description of the CIO/Equivalent in supporting the organisations environmental programs as the Systems Leaders

Describe the role of the Chief Information/Technology Officer or equivalent in supporting your organisations environmental program(s)? (select multiple responses)(%)

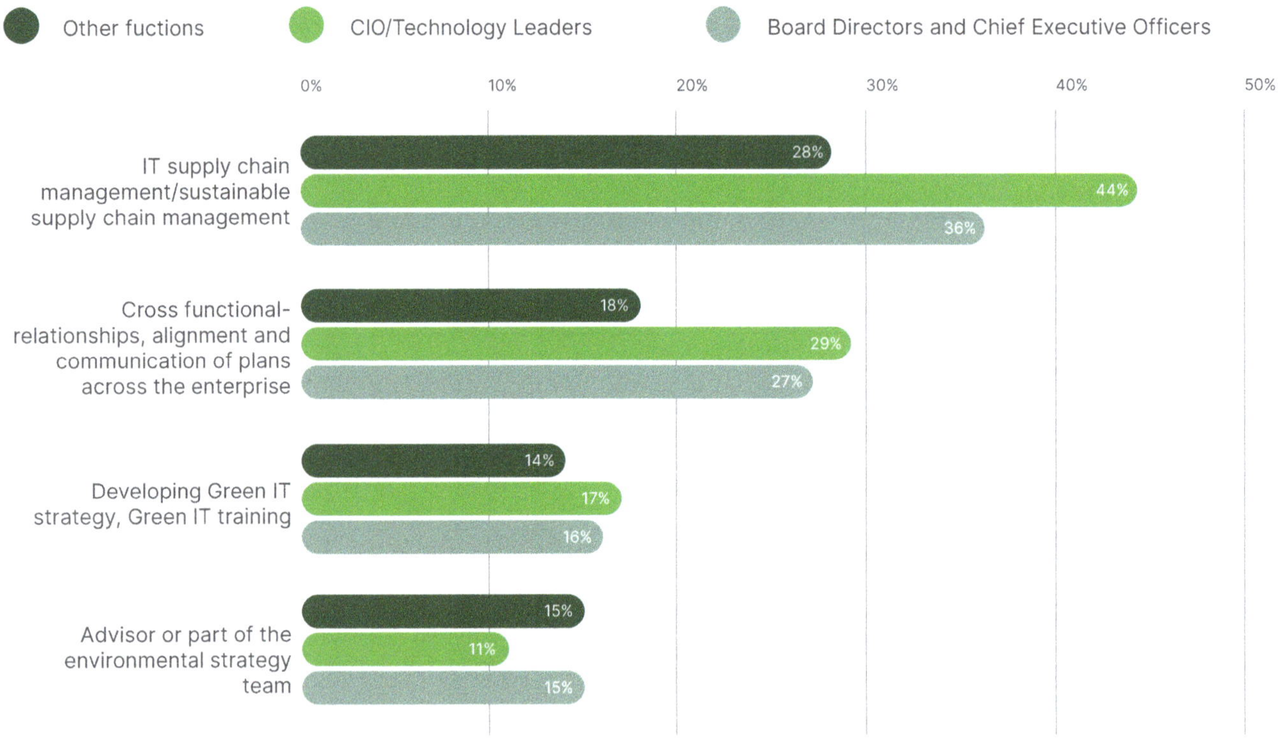

Source: *Australia 2030 Research*© Rocky Scopelliti (n=523); Other Functions include: Finance, Strategy, Human Resources, Sales, Marketing, Product Management, Risk, Legal, Compliance, Healthcare Care, Consultant.

Apart from IT supply chains, there is reasonable alignment between the board directors' and chief executives' perceptions of what the CIO should lead and the CIO's perception.

To accomplish the evolution of the role as proposed, both camps need to shift in tandem.

Large-scale initiatives such as those required to transform Australia's core economic systems and enabling services to meet industries environmental targets, are often driven, and supported by people who fit a certain profile. Those who can catalyse and empower collective action among others, rather than controlling or directing the action themselves.

These people are increasingly described as systems leaders (see Footnote 1). The systems leadership role and associated responsibilities (technological, governance, risk, financial performance, operations, values, policies etc.) requires the role of the CIO to move beyond a principally internal facing role, to a market facing role.

5.3. The CIO as the Environmental Enabler

When we compare responses from board directors and chief executives to CIOs or equivalents, and participants from all other functions, we also see significant opportunities to leverage the CIO in an environmental enabler role. For example, when it comes to the CIO as an enabler for an organisations environmental programs, we can again see a wide spread of results between the perceptions of board directors and chief executives, CIOs/Technology leaders vary widely in their perceptions of the role and its and characteristics (see **Exhibit 18**).

At an organisational environmental, operational, and program level only one in three board directors and chief executives and function leaders describe the role of the CIO as either an 'enabler for identifying, executing, and implementing environmental initiatives', 'developing new technologies and IT infrastructure', 'data management' or 'governance and compliance'.

This is quite disconnected from CIOs who describe their role here with much greater environmental characteristics.

It is critical to reconcile this discord as environmental programs are both data and technology intensive, and their sources are diverse and highly fragmented. Therefore, it is essential that the CIO is at the table of an organisation's environmental strategy, programs, and execution developments.

Exhibit 18: Description of the CIO/equivalent in supporting the organisation's environmental programs as the environmental enabler

Describe the role of the Chief Information/Technology Officer or equivalent in supporting your organisations environmental program(s)? (select multiple responses)(%)

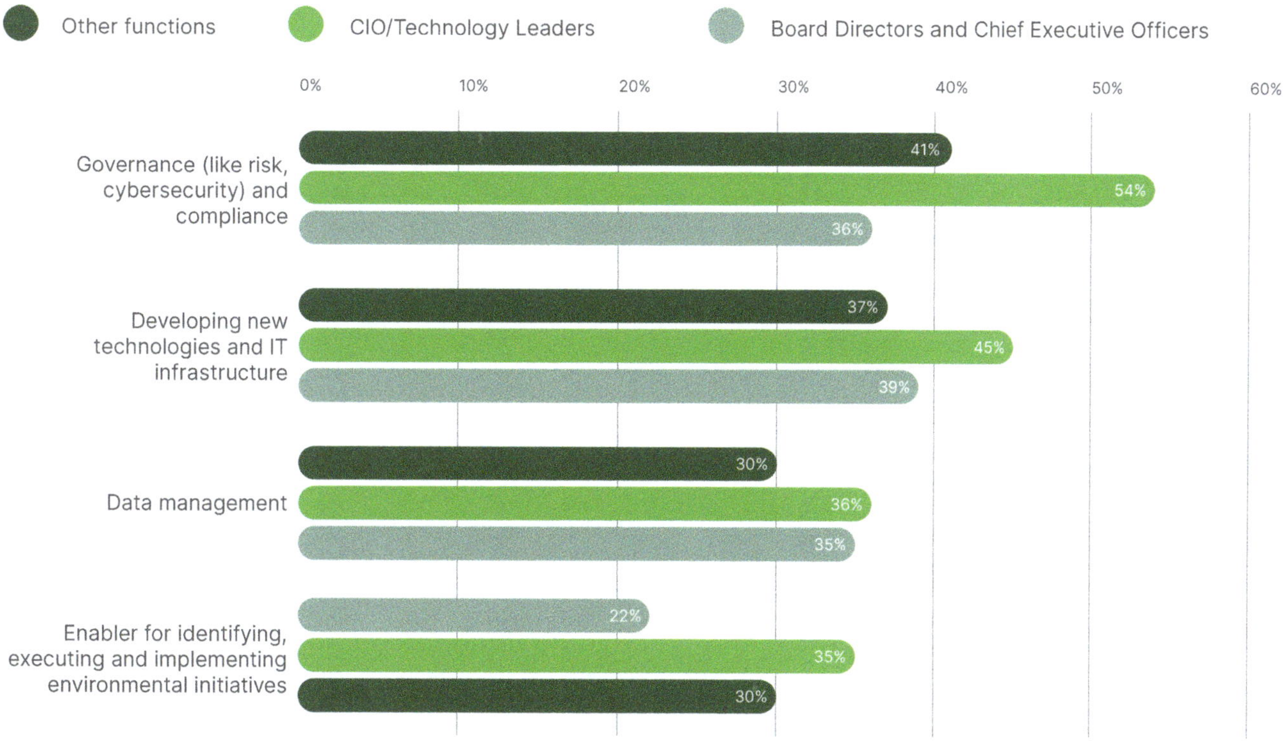

Source: *Australia 2030 Research*© Rocky Scopelliti (n=523). Other Functions include Finance, Strategy, Human Resources, Sales, Marketing, Product Management, Risk, Legal, Compliance, Healthcare Care, and Consultant.

The 'recapitalisation' of Australia's economy to net zero carbon emissions in less than 30 years will impact every industry, organisation, leader, worker, investor, policy maker and citizen. Unlocking the intrinsic characteristics of good systems leadership will be central to a successful transition for both core economic systems and enabling services organisations and the economy overall.

For any exponential strategic and operational transformation of Australia's industries and organisations, the CIO role is critical as an environmental enabler and systems leader, responsible for the organisation's environmental strategy and initiatives.

5.4. Section Summary/Key Takeaways

- Behind 'Climate Change', board directors and chief executives reported 'Skills, Reskilling, and Future-of-Work' as their next greatest concern about the decade ahead. Both these issues are inextricably linked, which opens the door for the next major evolution of the role of the CIO.
- CIOs today work alongside their C-suite peers and CEOs look to the CIO to drive innovation and revenue-generating ideas throughout the business.
- The CIO role needs to evolve to become the 'Chief Systems and Environmental Officer'. This evolution will be critical in orchestrating climate change transformation across Australian industries, government, and individual organisations this coming decade.
- However, the discord between how CIOs describe their systems leadership and environmental enabler characteristics when it comes to the organisation's environmental programs, and how board directors, chief executives and function leaders describe them, requires 'reconciliation'. This reconciliation is critical as environmental programs are both data and technology intensive and their sources are diverse and highly fragmented across the organisation.
- Large-scale initiatives such as those required to transform Australia's core economic systems and enabling services industries to the environmental targets, require systems leadership (technological, governance, risk, financial performance, operations, values, and policies etc.).
- For the exponential strategic and operational transformation of Australia's industries, the CIO's role is critical as the enabler and systems leader of each organisation's environmental strategy and initiatives.

6. Enterprise Technologies

"New technologies and approaches are merging the physical, digital, and biological worlds in ways that will fundamentally transform humankind. The extent to which that transformation is positive will depend on how we navigate the risks and opportunities that arise along the way."

Klaus Schwab, Chairman World Economic Forum

So far, the *Australia 2030 Research*© has explored the economic, strategic and leadership requirements. And importantly, their timing, for Australia to transform into a decarbonised economy. From here, the *Australia 2030 Research*© considers the role of existing and emerging enterprise technologies. And importantly, the necessary skills required to successfully deliver programs and initiatives that provide carbon reduction results.

As both an emitter and reducer of carbon emissions within an organisation, technology requires contrasting strategies to deliver its organisation's overall carbon emissions target. In doing so, an organisation can achieve short-to-midterm emissions reduction benefits from sourcing strategies, implementing key metrics across the enterprise, and deploying real time performance monitoring systems. The *Australia 2030 Research*© identified that, for many Australian organisations, these strategies, technologies, and real-time operational systems are rarely found.

6.1. Enterprise Technology – the emissions emitter and reducer

According to McKinsey & Company, although enterprise technology may not be a major emitter of global greenhouse gases, it is a significant contributor to carbon emissions in most major industries. It reports that technology is responsible for emitting an estimated 350 – 400 megatons of carbon dioxide equivalent gases (CO_2e) across a wide range of industries (see **Exhibit 19**).[12]

Their research also shows that enterprise technology carbon emissions are especially high for industries associated with enabling services like communications and media, banking, and insurance.

Exhibit 19: Global enterprise technology emissions (Upper Range Co_2e and percent share)

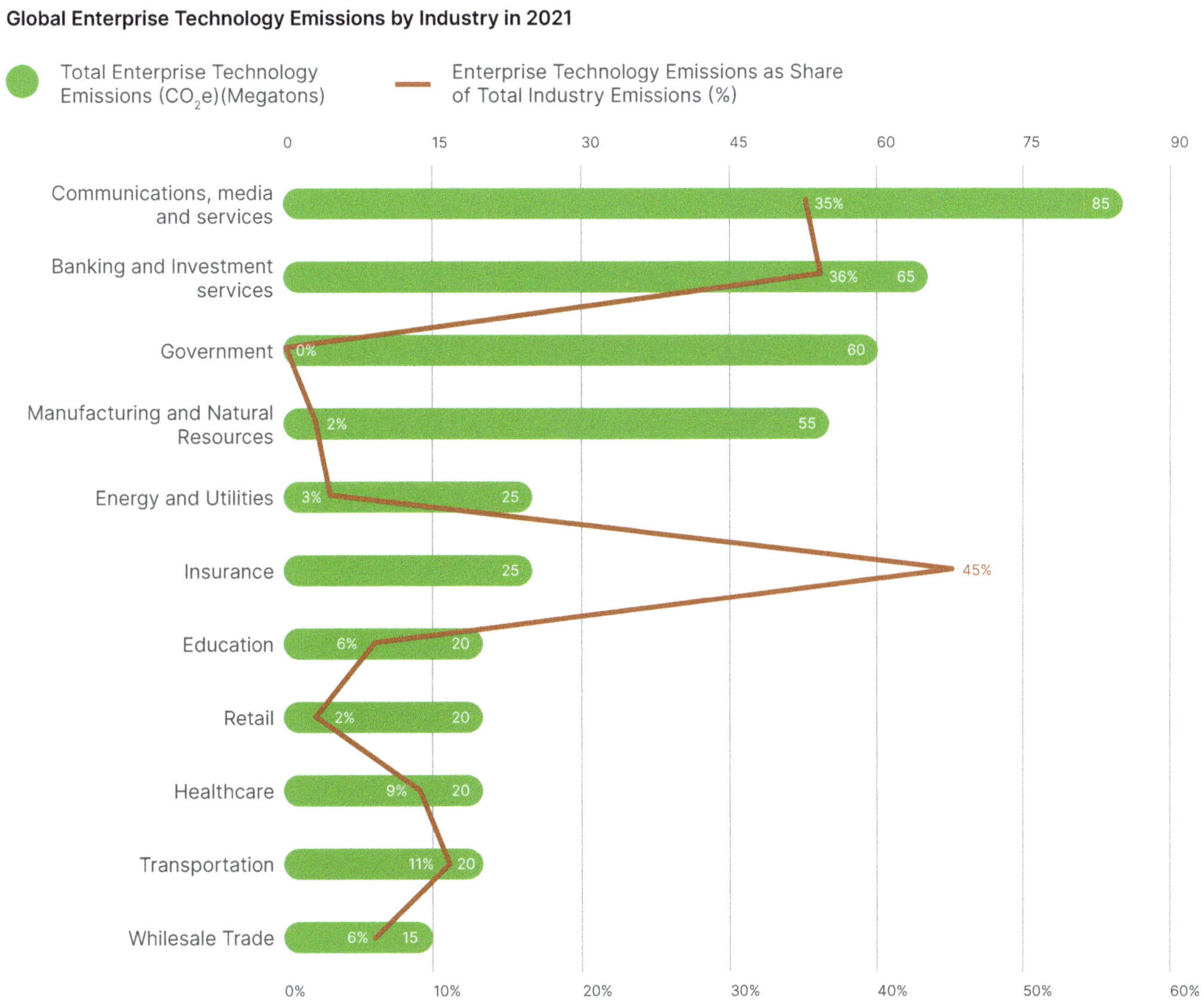

Data Source: *McKinsey Technolog*. Note upper CO_2e range illustrated. Lower range CO_2e approximately 5 – 10 megatons below the upper range.

This suggests that cutting the amount of carbon dioxide and equivalent gases generated by enterprise technology is a significant opportunity for companies seeking 'early wins' in cutting emissions. (Bear in mind though, that even those quick wins may take up to five years to realise due to technology and procurement lifecycles.)

There are several surprising facts about enterprise technology emissions. Firstly, the biggest enterprise technology carbon emitter is not the data centre as you might be led to think, but rather the end user devices, (which include laptops, tablets, smartphones, and printers etc.) Secondly, less than a third of Australian organisations are measuring the carbon emissions from their enterprise technologies. Lastly, those measuring carbon emissions, many are not measuring the right enterprise technology categories. Alarmingly, more than 52 percent of Australian professions reported that their organisations are not measuring any categories (see **Exhibit 20**).

At a sector level, the ***Australia 2030 Research***© found that 66 percent of those that reported their organisations were not measuring any enterprise technology carbon emissions, were from enabling services industries while just 34 percent were from industries associated with core economic systems. This suggests that organisations from high carbon emitting industries can benefit significantly from measuring their enterprise technology carbon emissions.

Exhibit 20: Global enterprise technology emissions by enterprise technologies (Upper Range Co_2e and percent share) and proportion of Australian companies measuring them

Global Enterprise Technology Emissions by Technology (CO_2e Megatons) and Proportion of Australian Companies Measuring Them

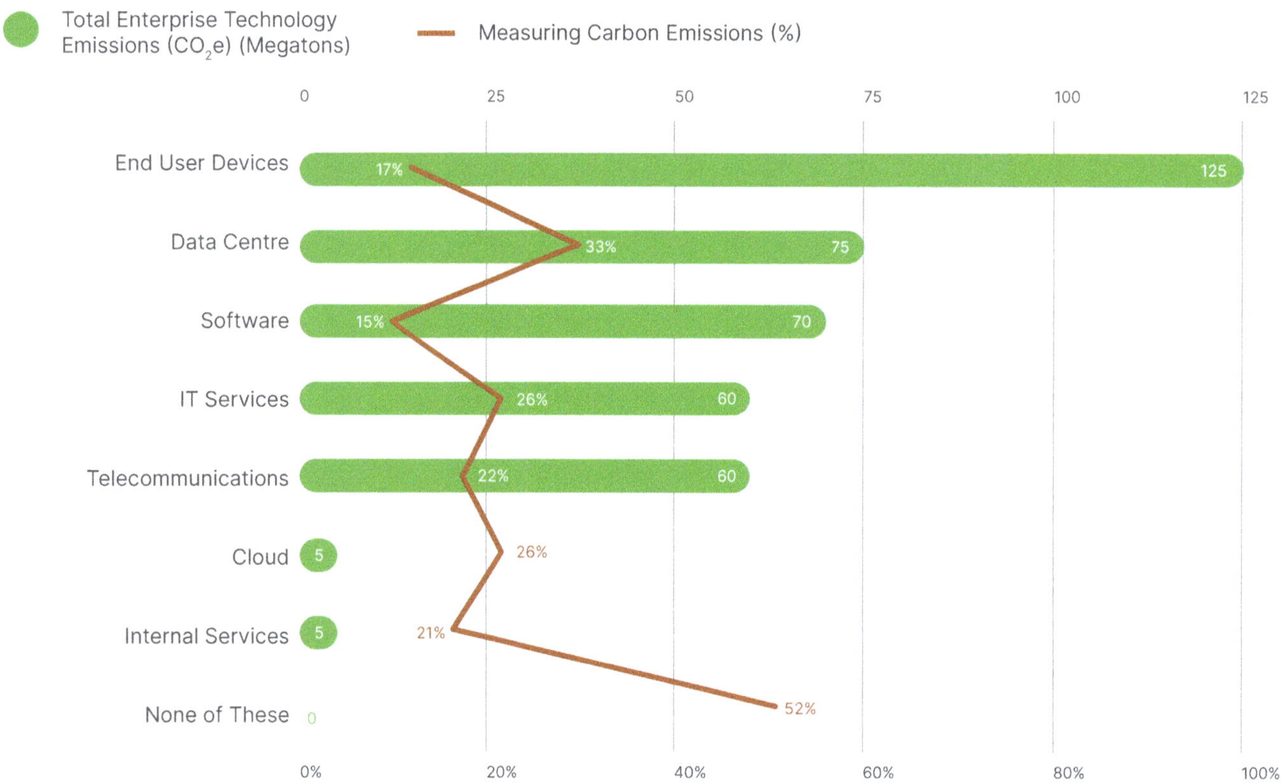

Source: *Australia 2030 Research*© Rocky Scopelliti (n=495); Data Source: *McKinsey Technology*; Note upper CO^2e range illustrated. Lower range CO_2e approximately 5 Megatons below the upper range.

If we look at just one of these technologies – cloud, we can see the significant opportunity measuring carbon emissions presents. A report by Deloitte in 2022 estimates that cloud-based solutions are, on average, five times more energy-efficient than on-premises data centres in the Asia Pacific.[13]

If all Australian organisations not yet on cloud storage converted into cloud users, there would be a potential savings of 4.5 million tonnes in carbon emissions a year – an equivalent of $1 billion in energy costs.[14]

6.2. Measuring Green Returns

Improving carbon emissions from existing enterprising technologies is not necessarily about more capital investments or operational expenditure on new technology.

But rather it can be optimising sourcing strategies, resource utilisation, lifecycle management, and asset management, increasing cloud migration, and the adoption of energy efficient technologies just to name a few.

It also requires a detailed understanding from technology suppliers of what materials are used in the manufacture of their technology products and services, where and how they are manufactured, how they are transported and much more.

The first step for Australian organisations is to establish enterprise-wide measurement systems across their enterprise technologies.

This will establish a baseline on which carbon emission reduction strategies can be deployed, measured, monitored and their benefits realised. This should go some way to mitigating the ***Australia 2030 Research©*** finding that 'clarity of benefits and unclear ROI' was the major concern for board directors and chief executives when it came to their organisation's environmental readiness this coming decade and that increasing revenue and profit were not seen as benefits realised.

McKinsey proposes that significant progress can be made by organisations when they adopt measures such as 'green returns'[15]. These include cost per ton of carbon saved and are monetised to include costs saved. These models can calculate emissions throughout the full life cycle, including production, distribution, transportation, repair and maintenance, and disposal.

Overtime, these measures and those under development by the ISSB, will become an integral part of an organisation's risk appetite statement. In this capacity, the Chief Risk Officer will then take responsibility for all climate change reporting, risk management strategies and operations, as well as supervising the organisation's risk mitigation and identification procedures.

Hence, effective measurement systems become critical to risk management. Recently, IT has become a big part of every business, and naturally, among others, the Chief Risk Officer needs to address the risks associated with data breaches and cyber security threats. Now climate change will be added to that list.

6.3. Technology Investments

Looking at the investments Australian organisations are making into their environmental programs can be broken down into two groups: technologies that are already mature and those that are emerging.

Many of the mature technologies attracting the most widespread investments relate to digital solutions, data and analytics capabilities, and workplace environments (see **Exhibit 21**).

Exhibit 21: Top technologies organisations are investing in for their environmental programs

What are the top technologies that your organisation is investing in for its environmental program(s)? (multiple responses)(%)

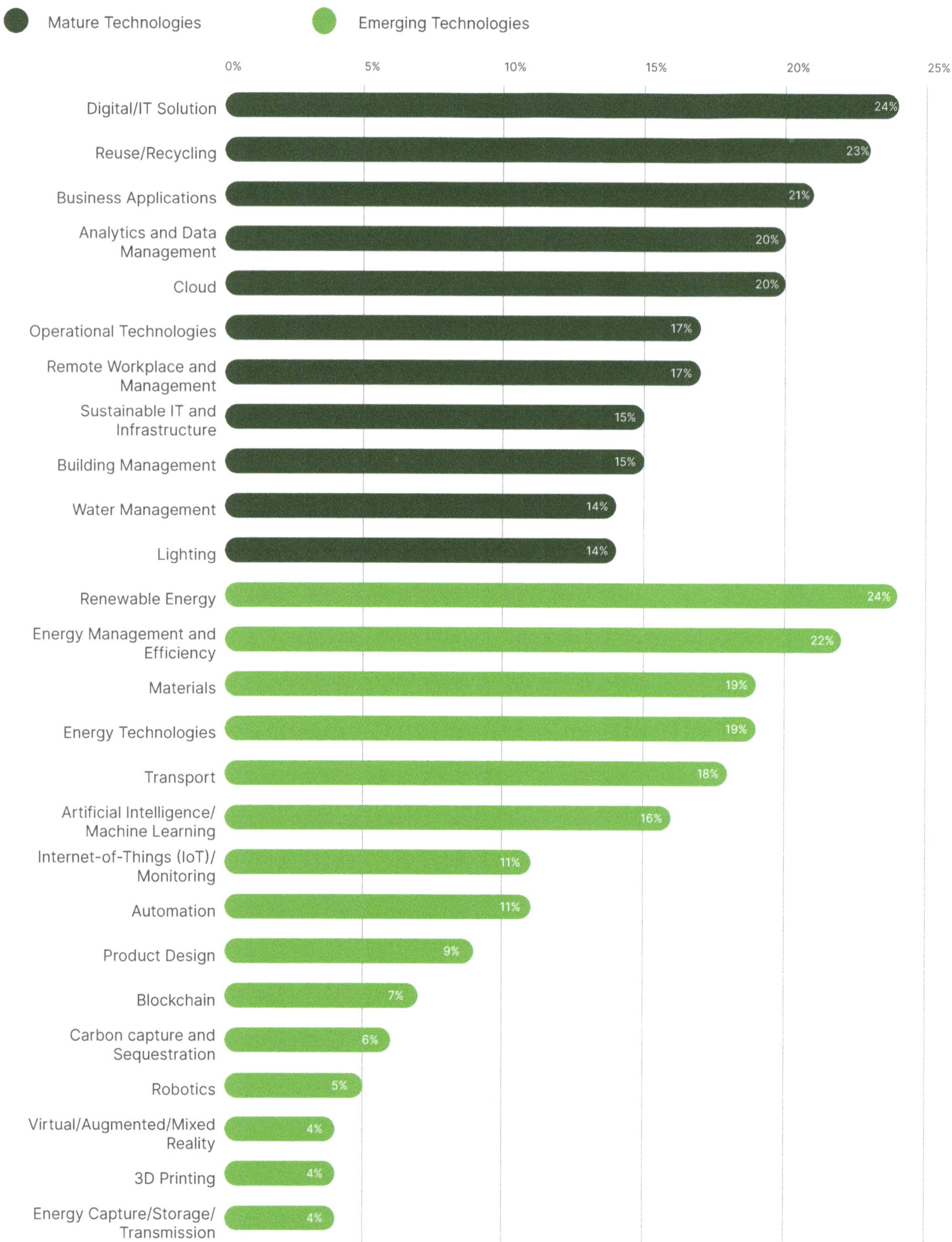

Source: *Australia 2030 Research*© Rocky Scopelliti (n=477).

Environmental programs are data intensive. They require lots of intelligent technology to gather, analyse and leverage the insights from operations and their value chains. Climate technologies are a necessary part of the solution. And leaders must respond to the urgency by adopting and scaling mature and emerging technologies at an unprecedented pace. This attracts risk.

Environmental technological transformations hinge on data and modelling transparency. The availability and management of accurate, verifiable data are both essential to achieving corporate carbon emissions metrics. Companies must both use tools to record, report, and act on sustainable business data within their operations and share them across their networks of suppliers and partners, (and eventually regulators) if they are to unlock the potential for industry decarbonisation at scale.

Emissions outside an organisations' own operations represent more than 80 percent of total emissions across most business sectors. End-to-end value chain transparency and sharing actual and verified emissions data, accounted down to the individual product and supplier level are all needed. Not estimates or averages. This was the point that the former Deputy Chair of the ACCC, Delia Rickard made (refer **Section 4.7**).

Moving from averages to actuals has enormous benefits. It allows organisations to embed sustainable business decisions into their business processes, from sourcing to product development. It allows them to invest in high yield decarbonisation activities in their supply chains, adhering to increasing regulatory requirements, and delivering more accurate corporate reporting are all elements of the ability to transform into a decarbonised economy.

We can also see emerging technologies attracting widespread investments, specifically in the use of intelligent automated solutions that improve energy efficiency, transport, processes, and buildings. And areas that help conserve resources such as water.

6.4. Technological and Scientific Developments

When it comes to looking at which technological or scientific developments that Australian leaders and professionals believed would have the greatest impact on environmental innovation in Australia in the coming decade, *Australia 2030 Research*© participants were asked to rank the following four options.

1. **Our Digital World:** By increasing our computing capacity, automation, connectivity, and speed through technologies such as Quantum Computing; Block Chain; Internet-of-Things; 5G Mobile Networks.
2. **Our Physical World:** Our cities becoming smarter, vehicles becoming autonomous, factories becoming automated and robotic through technologies such as AI; Robotics; Autonomous Vehicles; 3D Printing and Additive Manufacturing; Advanced Robotics; Material Science.
3. **Our Biological World:** Improving agricultural production, vertical farming, precision medicine for treatments through technologies including biotechnologies; neurotechnologies; virtual and augmented reality.
4. **Our Environmental World:** Renewable energies; technologies including energy capture, storage, and transmission; geo-engineering.

Participants were then asked which parts of their life they thought would significantly change due to those four areas of technological and scientific developments.

Approximately half the professionals believe that 'Our Environmental World' including renewable energies and technologies including energy capture, storage, and transmission, and geo-engineering would have the greatest impact on environmental innovation in Australia in the coming decade. This belief aligns with the $25 billion shown in **Exhibit 2** representing the largest estimated share of capital reallocation from emissions intensive, to low emissions assets. And perhaps represents an expectation of innovation in the energy system.

Interestingly, it further highlights the fact that one in two participants expect that these technological and scientific developments will have a significant impact on 'Jobs, career', 'skills development, studying, and learning' (see **Exhibit 22**).

Exhibit 22 :Technological and scientific developments over the coming decade

Which technological or scientific developments will have the greatest impact on environmental innovation in Australia in the coming decade? (%)

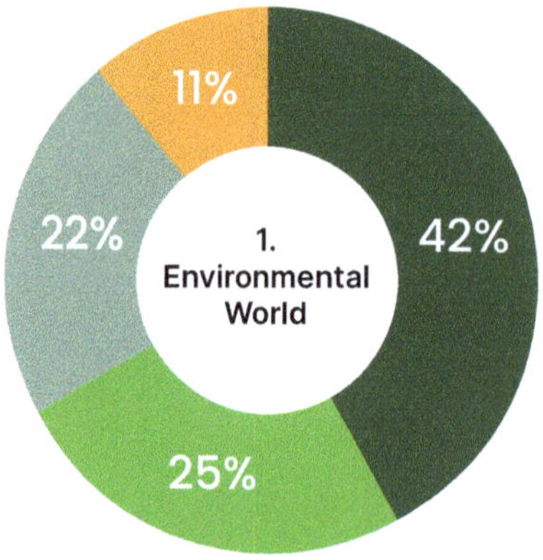

- Environmental World **42%**
- Physical World **25%**
- Digital World **22%**
- Biological World **11%**

What part of your life do you think is going to change significantly due to environmental technological and scientific developments over the coming decade? (%)

- Job, career **26%**
- Travelling, mobility **22%**
- Better physical and mental health **21%**
- Skills development, studying, learning **17%**
- Lifestyle e.g. work life balance **13%**
- Personal relationships, family, friends **1%**

Source: *Australia 2030 Research*© Rocky Scopelliti (n=506)

6.5. Section Summary/Key Takeaways

- The amount of carbon dioxide and equivalent gases generated by enterprise technology presents a significant opportunity for 'quick wins'. However, benefits may take up to five years to realise due to technology and procurement lifecycles.
- Less than a third of Australian organisations were reported to be measuring the carbon emissions from their enterprise technologies. Of those that are measuring, many are not measuring the right enterprise technology categories, and alarmingly, more than half the board directors and chief executives reported their organisations are not measuring any of their enterprise technologies.
- This again explains why 'clarity of benefits and unclear ROI' was the major concern for board directors and chief executives when it came to their organisation's environmental readiness this coming decade and the link to them not seeing revenue generation and net profit increases.

- The first step for Australian organisations is to establish enterprise-wide measurement systems across their enterprise technologies so a baseline from which carbon emission reduction strategies can then be deployed and monitored and the benefits measured.
- As measures become an integral part of an organisation's risk appetite statement, the Chief Risk Officer will then become responsible for all climate change risk management strategies and operations, as well as supervising the organisation's risk mitigation and identification procedures.
- Improving carbon emissions from existing enterprising technologies is not necessarily just about spending more on new technology – but rather, also optimising sourcing strategies, resource utilisation, lifecycle management, and asset management, increasing cloud migration, and the adoption of energy efficient technologies, just to name a few.
- More than half of all organisations were reported to not be measuring any carbon emissions from their enterprise technology. Most of these were from enabling services industries which stand to benefit significantly from measuring their enterprise technology carbon emissions.
- Climate technologies are a necessary part of the solution, and leaders must respond to the urgency by adopting and scaling mature and emerging technologies at an unprecedented pace.
- The mature technologies attracting investment today include digital solutions, data and analytics capabilities, and workplace environments. Emerging technologies attracting investments include intelligent automated solutions that improve energy efficiency, transport, processes, and buildings, and those that conserve resources such as water.
- The belief that renewable energies, technologies including energy capture, storage, and transmission as well as geo-engineering would have the greatest impact on environmental innovation in Australia in the coming decade. This aligns with the significance of the capital reallocation in the energy system.
- Australian professionals expect that their 'jobs career', 'skills development, studying, and learning' will be the most impacted by these technological and scientific developments.

7. The Reskilling Imperative

"We're currently 186,000 jobs short of the 1.2 million technology jobs we'll need in 2030."

Scott Farquhar, Atlassian Co-Founder

The *Australia 2030 Research*© has shown that the lack of 'skills and reskilling' is board directors and chief executives second biggest concern when it comes to their organisation's strategic environmental readiness over the coming decade.

The *Australia 2030 Research*© further showed that Australian professionals reported that their 'jobs career', or 'skills development, studying and learning' will be the areas of their life most affected by new technological and scientific developments.

Therefore, in this section the *Australia 2030 Research*© considers the intergenerational demographic disruption Australia faces in its supply of workforces as it looks to 2050.

It presents the case for organisations to realise a demographic dividend. They can do so by unlocking Australian workforces' intrinsic cultural capacity to adapt. This ability will be critical to realise the benefits from decarbonising their organisations.

7.1. Australia's Ageing Population

For Australia the effects of an ageing population will be felt more over this coming decade, than in the past. This is due to the impact of the Baby Boomers (born broadly between 1946-1964) retiring. This massive ageing trend has implications for nearly all aspects of society, including labour and financial markets, healthcare, the demand and supply of goods and services, such as housing, transportation, and social protection, as well as family structures and their intergenerational ties.

Australia's ageing population will have a particularly profound effect on the available labour market and workforces. The Organisation for Economic Co-operation and Development (OECD) applies a potential support ratio measure, defined here as the number of people of working age (25 to 64 years) per person aged 65 years or over to consider the impact of aging populations. This helps us observe the status of a nation or regions.

For Australia in 2020, the ratio has 3.2 persons aged 25 to 64 for each person aged 65 or over declining to 2.3 in 2050. Virtually one working age person in the next 30 years.

This declining ratio also exists in other developed nations such as the United States, Canada, Europe, Japan (see **Exhibit 23**).

Exhibit 23: Forecast potential support ratio number of people of working age (25 to 64 years) per person aged 65 years or over

Estimated and projected support ratio, 2020-2050 (ratio persons aged 25-64 per person aged 65 or over)

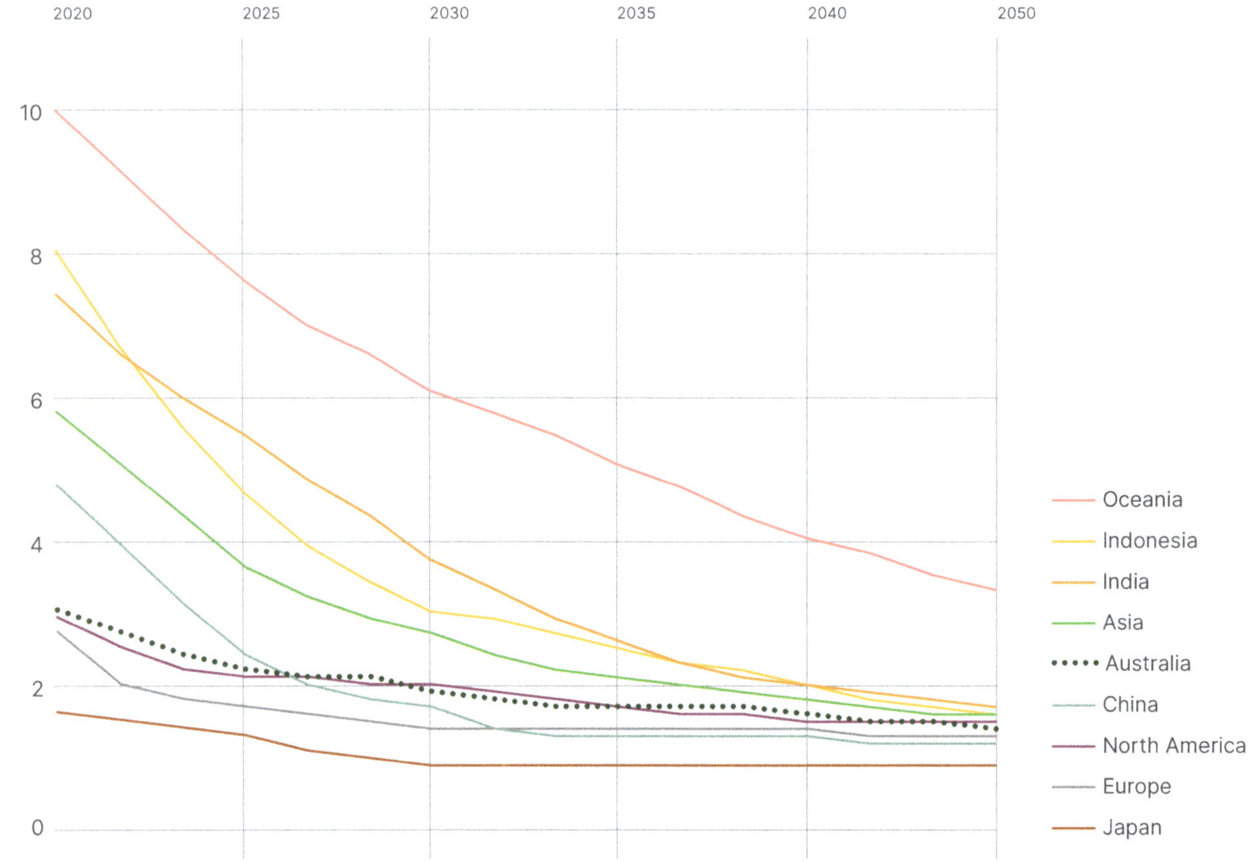

Data Source: United Nations, Department of Economic and Social Affairs, Population Division (2019). World Population Prospects 2019

The current and projected profile means that Australia will need to identify more effective ways to leverage its increasingly scarce human capital resources, more productively, efficiently, and dynamically.

7.2. Adaptive Capacity

The *Australia 2030 Research*© found that 43 percent of professionals across all demographic and professional groups ranked 'job, career, skills development, researching, and learning' as number one when asked: 'What part of your life do you think is going to change significantly due to environmental technological and scientific developments over the coming decade?'. There is a strong appetite for reskilling.

This is a critical cultural finding because core economic systems industries employ approximately half of Australia's workforce with the other half employed by enabling services.

In the long-term however, Australia, like others with a similar outlook to 2050, will need to identify new ways to skill and reskill workforces that have grown up digitally, are purpose driven and crave development.

According to projections by the United Nations[16], the number of older persons globally is growing faster than any other age group globally. In contrast, over the same period, the number of people 24 years old or younger will grow a mere 11 percent, and the number of people aged 25–59 will grow by 62 percent.

These projections underline the significant representation Millennials will continue to have in the workforce through to 2050, as those Millennials born around 1980 reach 70 years of age.

Importantly, by 2050, there will be another two generations who have either only just now entered the workforce (Generation Z) or are very young or still being born (Generation Alpha, born broadly between 2010-mid 2020s).

In a recent global study, only 33 percent of technology jobs worldwide were reported to be filled by the necessary skilled labour[17].

This is one of the biggest concerns of global business leaders. Half the organisations surveyed say that the digital divide is widening. And 54 percent say they have lost competitive advantage due to talent shortages.

The World Economic Forum estimates that the world will create 150 million new technology jobs over the next five years. And by 2030, 77 percent of jobs will require digital skills[18].

Australia has a remarkable history of innovation, but we need to invest in the next generation of high value-adding jobs to sustain our strengths.

As an advanced economy, there is a greater opportunity for Australia to invest in the high value-added pre and post-production stages of manufacturing. But the digital and data literacy of our workforces is critical.

As Treasury highlighted in its recently published *Jobs + Skills Summit Issues Paper*[19], making the most of the opportunities created by these big structural transformations and creating the next generation of high value-adding jobs requires having the right skill mix for the long term.

At the same time, we need to ensure we are taking steps to build a more agile, skilled, inclusive, and resilient labour force for the future.

Treasury suggested that for Australia to maximise opportunities from new, growing and strategically important industries including renewable energy, the digital and care economies will be critical to boosting Australia's productivity, sustaining full employment, and ensuring Australia's cities and regions thrive.

Building a more agile, diverse, and skilled labour force will also be critical in an era of rapid structural change.

7.3. The Demographic Dividend

Young adults and the youth of today are critical to Australia's 2050 environmental targets.

A demographic dividend refers to the growth in an economy that results from a change in the age structure of a country's population. This is achieved in two ways.

Firstly, through investment in economic development and family welfare leading to growth in per capita income.

And secondly through the accumulation of assets that become invested domestically, or internationally, resulting in rising national income.

While often applied to describe the economic development opportunities in emerging markets with a higher proportion of working age people relative to their aged population as shown in **Exhibit 23**, I believe that opportunities for creating this dividend in Australia do exist. But they need to be applied in a transformative way.

When asked: 'Do you feel positive or negative about the speed of technological and scientific change to support Australia's environmental goals over the coming decade?', the *Australia 2030 Research*© found that 52 percent of Millennial and Generation Z participants were positive. But only 32 percent from other demographic categories were positive (see **Exhibit 24**).

Exhibit 24: Feel positive about the speed of technological and scientific change to support Australia's environmental goals over the coming decade

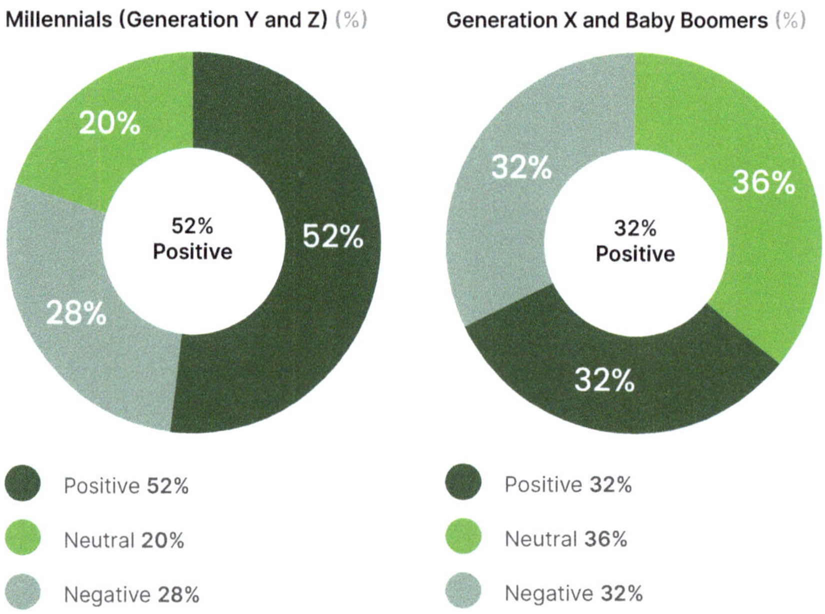

Source: *Australia 2030 Research*© Rocky Scopelliti (n=520)

The Millennial, Generation Z and Generation Alpha's optimism about the speed of technological and scientific change to support Australia's environmental goals this coming decade likely stems from the fact they grew up digitally connected. They have had to adapt to a world changing much faster than generations before them.

> **As workforces, they are our most technologically proficient and adaptable demographics. They will primarily be responsible for delivering Australia's carbon emissions targets in both 2030 and 2050. They live in real-time and adapt exponentially. Critical to transforming to a low carbon emissions economy.**

7.4. Reskilling Workforces

Given the government's intention to grow the digital economy and reach 1.2 million tech jobs by 2030, technology can provide solutions to some of the country's most pressing needs. By reducing friction in the job search process, and enhancing digital skills training, digital technologies can boost labour productivity and estimated to create up to an additional $25 billion in Gross Domestic Product (GDP) per year[20].

Over the years, the digital economy has created a steady stream of new job opportunities for Australians. Yet it still faces a shortage of skilled workers in many areas of the economy. The talent supply to fill these technology-focused roles has been unable to keep up with demand. This has hindered the nation's ability to benefit from the fast-growing digital economy.

However, if Australia can address the digital skills gap, it will boost labour productivity and turbo charge the economy. For example, it is estimated that it could help businesses in the technology, media and telecommunications industries grow by $10 billion by 2025[21] and create up to an additional $25 billion in Gross Domestic Product (GDP) per year.

Based on data from the Australian Department of Education, an average of 6,500 students graduated with IT degrees each year.[22] According to Deloitte Access Economics, this fails to meet the yearly demand for the additional 156,000 more digital technology workers Australia will need by 2025.

According to the Climate Council, clean energy industries are facing an especially acute skills shortage, with three in four solar companies finding it difficult to recruit electricians with adequate experience[23]. They report that investing in skills training for renewable energy could create an extra 30,000 jobs in Australia as we shift to net-zero emissions.

Skilled workers are needed to construct renewable energy infrastructure, to work on solar farms, wind farms, batteries, transmission lines and pumped hydro, as well as in energy efficiency and clean transport.

But this is just one of the industries associated with the four core economic systems that are facing skills shortages, together with those in enabling services. Australia will need to be disruptive and ingenious to find ways to reskill its workforces and unlock the intrinsic value of its human capital.

Nevertheless, Australia can continue strengthening its economic growth by developing a capable and adaptable workforce through regular up-skilling and reskilling.

The three ways Australia can improve the labour market's digital readiness are through:

1. Personalised learning to enhance the quality of education and increase the supply of technologically skilled labour.
2. Digital job platforms and career matching services that are key to reducing labour market frictions.
3. Retraining programs that enable workers to equip themselves with new skills for emerging industries and roles.

Organisations are increasingly taking much greater control of their skills development requirements and benefit from both unlocking intrinsic value and futureproofing their workforces. More than ever, organisations that support upskilling employees have a clear competitive advantage in an increasingly tight labour market and rapidly changing business environment. To achieve this, some are now collaborating with academic institutions for this strategy. One such institution is the University of Technology Sydney (UTS). Its offers expertise in agile learning for a world in constant flux through its PLUS UTS Business Futures (see **Exhibit 25**).

Exhibit 25: Case Study: University of Technology Sydney's PLUS UTS Business Futures

PLUS UTS Business Futures, designs and delivers evidence-based and engaging learning experiences that create an immediate impact for clients by fostering a shift in mindsets, behaviours, and practices.

This is achieved by translating latest academic research and leading industry practice into a set of activities, tools, and techniques that learners take away and apply in their work practices.

By helping workers tackle tasks of increasing complexity that are directly relevant to their role and experience, these upskilling programs increase workers' confidence and ability to adapt to new challenges and opportunities. The human-centered, agile approach used by the team to design and deliver programs ensures that program outcomes meet both the individual and organisational upskilling needs and requirements. And contribute to improved performance on the job and retain talent.

Organisations that have leveraged these upskilling programs include John Holland Infrastructure, NSW Public Service Commission Leadership Academy, NSW Department of Planning and Environment, Commonwealth Bank of Australia, Reserve Bank of Australia, CPA Australia, and many more.

PDNSW called on PLUS UTS Business Futures to build its talent and capability to better respond to the increasingly complex and challenging environment in which it operates.

"Delivering great outcomes for the people of NSW requires effective critical thinking and problem solving. PLUS UTS Business Futures were able to develop a program tailored to our unique needs and requirements. The content was engaging and directly applicable to our role delivering better use of government land and property."

Leon Walker, Deputy Secretary – Homes, Property and Development
Department of Planning and Environment

7.5. Section Summary/Key Takeaways

- For Australia, the effects of an ageing population will be felt more over this coming decade than ever before due to the retirement of the Baby Boomer generation. This will place significant strain on the labour market and add to the skills shortage over the long term.
- The appetite for reskilling is very strong among Australian workforces. All demographic groups expect 'job, career, skills development, researching, and learning' to be significant due to environmental technological and scientific developments.
- Opportunities for creating a demographic dividend do exist for Australia, if applied in a transformative way. This is evident by most of the younger population that feel positive about the speed of technological and scientific change required to support Australia's environmental goals over the coming decade.
- To unlock the intrinsic value, and future-proof workforces to realise this demographic dividend, organisations are increasingly taking much greater control of their skills development requirements.
- By helping workers tackle tasks of increasing complexity that are directly relevant to their role and experience, upskilling programs such as those offered by the University of Technology Sydney (UTS) PLUS UTS Business Futures, increase workers' confidence and ability to adapt to new challenges and opportunities. The human-centered, agile approach used by the team to design and deliver programs ensures that program outcomes meet both the individual and organisational upskilling needs and requirements, contributing to improved performance on the job and the retention of talent.

8. Industry Digitalisation and Distributed Emerging Technologies

To recap, Australia's board directors and chief executives' primary concerns about their organisations' environmental strategic readiness over the coming decade as identified by our *Australia 2030 Research*© (see **Exhibit 8**) are:

1. 'Clarity of benefits and unclear return on investment'.
2. 'Lack of Skills and Reskilling'.
3. 'Under investment in the necessary enabling technologies'.

Because the *Australia 2030 Research*© further shows that environmental programs are both data and technology intensive with diverse and highly fragmented sources, the CIOs role is critical. Chief Information Officers are the enablers and systems leaders of their organisation's environmental strategy and initiatives and have the capacity to exponentially transform Australia's industries and organisations strategically and operationally.

This final and comprehensive section of our *Australia 2030 Research*© outlines the digitisation of some industries and the technological opportunities they provide. It presents examples and detailed definitions of the innovative opportunities emerging technologies could deliver to provide the exponential change required for organisations' environmental programs.

8.1. Digitalisation and Carbon Emissions

Analysis by Accenture, in collaboration with the World Economic Forum, shows that digital technologies, if scaled across industries, could deliver up to 20 percent of the 2050 reduction needed to hit the net zero target.[24]

Digital technologies, such as communication systems, geospatial data analytics and drones, can significantly minimise the costs of future natural disasters in Australia. For example, improved communication technologies can reduce emergency response times and improve the efficiency of disaster response and support teams. Likewise, AI can help enhance accuracy when detecting and forecasting future natural disasters. While geospatial data analytics and drones allow rescue teams to map out areas of damage in real-time and search larger areas for more efficient aid delivery (see **Exhibit 26**).

Exhibit 26: Case Study: New ANU-Optus partnership to build national defence system against catastrophic bushfires[25]

Optus and the Australian National University (ANU) have joined forces to develop a revolutionary national system that aims to detect bushfires early and extinguish them in minutes.

The ANU-Optus Bushfire Research Centre of Excellence will undertake advanced research and develop hi-tech solutions to predict, identify and extinguish blazes before they become deadly.

The ambitious program will run until 2024. In the short-term, experts from Optus and ANU will work together to develop an autonomous ground-based and aerial fire detection system.

By 2022, the program proposes to launch a constellation of satellites, managed by ANU, to complement the fire detection system. The program will look to be augmented by a geostationary satellite to help spot and track fires, as well as deploy extinguishing technologies.

It will investigate how to use existing and new technologies including infra-red cameras, drones, robotics, and satellites. It will also harness expertise and research in space, communications, computer vision, sensing systems, defence, data analytics and bushfire science.

Bushfires are expected to cost the nation at least $30 billion over the next three decades. Recent modelling from ANU shows investment in early bushfire detection could save Australia $8.2 billion over the next 30 years.

In 2030, the application of technologies in disaster response is expected to mitigate up to $4.5 billion in annual damage to property, crops, and livestock. More importantly, digital technologies can protect Australians from up to 591 life years[26] lost due to deaths and injuries caused by natural disasters in 2030[27].

Emerging technological and scientific advancements are occurring at exponential rates. Australia's peak body for innovation technology, the Australian Information Industry Association (AIIA), published a White Paper in 2021 titled *'Growing Globally Competitive Industries – Powered by Australia's innovative technology'*[28]. This paper provides a range of recommendations to Government, including to:

- Take a lead in accelerating the digitisation of Australian industry through creating hubs and incentives.
- Play a leading role in coordinating Industry, Government and Technology to initiate a new era of technology innovation in Australia.
- Ensure standards and interoperability requirements are in place to maximise the opportunity that technology presents to Australia.
- Ensure the new generation of Australian technology innovators are supported to thrive, not just survive, through a strong shift from research driven agenda, to one of commercialisation.

The following case studies highlight where digitisation and innovation will be critical across key industries in Australia to improve our global competitiveness in a decarbonised world economy. We also look at some of the technological developments that are advancing rapidly and will impact industries associated with both core economic systems and enabling services.

8.2. Agriculture

Agricultural technology, or AgTech, is predicted to become Australia's next $100 billion industry by 2030[29]. The sector is fast becoming a vital part of the Australian economy and a growing source of both income and investment. Thriving AgTech ecosystems, world-class research institutions and high-quality produce, make Australia an ideal environment for AgTech innovation.

Global warming and diminishing natural resources present obstacles to meeting changing demand. So too does the rising cost of energy, labour, and production inputs. These challenges have created the perfect environment for AgTech to flourish.

Recently, the opportunities in agricultural and food innovation global markets continues to grow, attracting greater international focus and investment. Australia has not yet managed to attract a significant proportion of this investment. This is despite its competitive, if somewhat immature, AgTech Start-Up ecosystem. Of the US$16.9 billion invested globally in 2018, Australia only attracted US$29 million[30].

Australian agriculture accounts for:

- 55 percent of Australian land use (427 million hectares, excluding timber production in December 2020) and 25 percent of water extractions (3,113 gigalitres used by agriculture in 2018-19).
- 11 percent of goods and services exports in 2019-20.
- 1.9 percent of value-added (GDP) and 2.6 percent of employment in 2019-20.

Given slowing productivity and smaller farms less profitable on average than larger ones, there is a real opportunity to leverage technology innovation. With the right investments and policies, Australia's AgTech industries could achieve similar levels of productivity and profitability to the world's larger counterparts.

Specific opportunities in AgTech for Australia to address productivity in the agriculture industry include:

- Robotics and automation.
- Supply chain traceability including use of blockchain technology.
- Synthetic biology and creation of new proteins opportunities.
- Vertical farming (see **Exhibit 27**).

Exhibit 27: Case Study: Sustainable, reliable food production with modern, vertical farming[31]

Helping to support efficient and sustainable modern farmers and feeding the world in a 'smarter way' are complimentary business and social impacts objectives for the team at InvertiGro.

The Sydney-based company – named as a finalist in the 2022 InnovationAus Awards for Excellence in the Food and Agritech category – designs, develops, builds, and supports Indoor Vertical Farming solutions that help make urban agriculture profitable and sustainable.

The technology enables businesses – wherever they sit in the food supply chain (from on-farm to in-store) or across the planet, to become efficient modern farmers.

The product, technology, and service offering include patent-protected hardware, a bespoke set of smart farm management software, and a range of support services that enable customers to become farmers rapidly and affordably, anywhere across the globe.

8.3. Health

Digital technologies have significantly advanced health services in Australia. Patients, health professionals and funders – as seen in the accelerating uptake of telehealth, virtual care, and e-prescriptions. However, for an advanced economy Australia is still in its infancy in terms of digital health development and deployment when compared with more advanced countries.

The World Index of Healthcare Innovation ranks Australia at 26 out of 31 countries for Science and Technology, citing restrictions on the use of new technologies as a major barrier to innovation[32].

A new focus on the digitisation of health is warranted given that the health industry is growing faster than any other in Australia, with employment expected to increase by 12 percent or 200,000 between 2020 and 2024[33].

Technology, automation, and Artificial Intelligence (AI) are driving advanced changes in health services across the world, for the following reasons.

- **Greater Access** – Developing, promoting, and making available various channels through digital solutions and platforms give more of the community access to health services via private and public health providers.
- **Greater Efficiency** – Digital transformations are being deployed in small and large health organisations, including hospitals. Their efficiencies are streamlining processes, avoiding replication, and reducing the cost of healthcare through agile and manageable implementations.
- **Mobility** – The public are more connected now than they have ever been. The technology underpinning smartphones allows them and service providers to connect on many levels, anywhere, and at any time of the day.
- **Assurance** – Using digital technologies to put seamless, traceable, auditable, health services in place provides solid assurance.

In the latest 2022 Australian Federal Budget, the Government made provision to fund a National Health and Climate Strategy. It allocated AUD$3.4 million over four years (and $0.7 million per year ongoing) to develop a strategy for and establish the National Health Sustainability and Climate Unit. From this unit, the Australian Medical Association (AMA) and Drug Enforcement Agency (DEA) want to see all health departments across Australia working together to reduce emissions and elevate sustainable practices in healthcare.

As Australia's healthcare system contributes up to 7 percent of Australia's carbon emissions, there is growing support from the health profession to significantly reduce its carbon footprint and be a leader in environmentally sustainable healthcare.

The AMA and DEA have an interim goal of an 80 percent reduction by 2030 — a commitment that the AMA and DEA lead. This won't be limited to decreasing energy use and waste. It will extend to better procurement decisions, improved infrastructure and planning, public and preventive health care, and innovative care pathways.

8.4. Manufacturing

Although manufacturing peaked at 25 percent of Australian GDP in the 1960s, and now represents less than 10 percent of GDP due to global market forces, it is still a vitally important part of the Australian economy. And one which stands to substantially benefit from current advances in technology.

While considering economies like Germany's, that have long embraced Industry 4.0 and have boosted productivity by embedding technology in its manufacturing and traditional industries, Australia has lagged. According to the May 2020 Australian Manufacturing Performance Report, Australia's manufacturing industry currently contributes around $100+ billion to Australian GDP annually and employs around 900,000 Australians. Manufacturing also contributes a high proportion of business expenditure on research and development (R&D) at 26.4 percent.[34]

The critical importance of manufacturing for Australia was underlined in the October 2020 Australian Federal Government budget. It announced that $1.5 billion of new funding would be invested over four years in the Modern Manufacturing Strategy (MMS)[35] to make Australian manufacturers more competitive, resilient, and able to scale to take on the world.

The MMS has its focus on lifting Australia's capabilities in key sectors including:

- Resources Technology and Critical Minerals Processing
- Food and Beverage
- Medical Products
- Recycling and Clean Energy
- Defence
- Space

The key common technology enablers across these industry sectors include:

- Cyber security
- Artificial intelligence
- Digital twinning and digital engineering
- Big data and analytics
- Automation
- Cloud infrastructure
- Collaboration tools
- Blockchain
- 5th Generation Mobile Networks (5G) and Edge Computing
- Internet-of-Things (IoT)

8.5. Digital Twins

A digital twin is a digital representation of a physical asset, process, or system, that provides information to allow its users to understand and model its performance. It is a living 'thing' – like an organism, enriched by continuously updating itself from multiple sources to mirror working conditions and represent operations near real-time.

A digital twin ecosystem for Australia would bring together a broad range of government, industry, construction, manufacturing, transport, and utilities organisations. The Australian Information Industry Association (AIIA) – Australia's peak industry body in the information technology sector, is encouraging governments at all levels to engage in strategic and long-term planning. It recommends ramping up investments in cutting-edge digital infrastructure to ensure better climate change influenced preparedness and coordination of flood and fire intelligence. This will ultimately save lives and property and is sufficiently intelligent to predict potential issues and recommend critical preventative measures.

In a media release AIIA CEO Simon Bush said: "The use of real-time, visual, simulation-capable digital twins of natural systems can provide authorities with a vital tool for smart flood and bushfire preparedness, especially when leveraged with Internet of Things, point solutions, flood sensors and Artificial Intelligence.

"In changing landscapes and climates, analogue flood maps are liable to become out of date. The dynamism and agility of the digital twin model, when paired with rich streams of data, could be a game-changer for assessing flood-prone areas."[36] (see **Exhibit 28**).

Exhibit 28: Case study: The NSW Spatial Digital Twin[37]

The NSW Spatial Digital Twin, a NSW Government program led by the Department of Customer Service is a useful case study. It provided an extensive range of new 3D foundation spatial datasets with geographic coverage at its launch in March 2020. These included the Western Sydney City Deal partnership that brought together federal, state, and local government under a set of 38 commitments. The Spatial Digital Twin was part of the project's commitment to coordinating information that supports planning and broader liveability targets.

Data has become its own type of asset. And the NSW Spatial Digital Twin is a data infrastructure that will benefit private industries, government, and the public into the future. The project sets a standard for how to make and use urban, or city scale digital twin data platforms for data discovery, sharing, and visualisation. As well as applications' domains, such as integrated planning.

At launch, the NSW Spatial Digital Twin was the largest 3D modelling project ever undertaken. Its final product – the interactive tool – captured 22 million trees with height and canopy attributes, almost 20,000km of 3D roads, and 7,000 3D strata plans with 546,206 buildings at under 5cm pixel resolution across the greater Western Sydney region[38]. And this coverage continues to grow.

Award-winning Aerometrex, an Australian 3D modelling expert was key to this endeavour. It captured and processed the necessary 3D data to underpin the NSW Spatial Digital Twin, spending several months capturing, processing, and combining the imagery.

Its open-source, publicly accessible browser-based web portal makes the NSW Spatial Digital Twin remarkable. Developed in partnership with CSIRO Data61, this foundational information source brings compounding efficiency benefits to government, private industry, and the public. All these entities can refer to a central and consistent data source.

Building a successful Digital Twin, whether it is a geospatial, manufacturing, design or service twin, generally has a data gathering phase. This includes operational, sensor and base knowledge data. This is followed by a 'thinking phase', including cognitive, reasoning, optimising, and machine-to-machine simulations. Culminating in an executing phase where the digital twin constructs, informs, acts, and edge controls. For each of these phases within the continuum there are varied necessary skills including:

- Digital Twin –awareness and understanding.
- Data literacy skills.
- Information modelling – metadata, integration, and attribution.
- Digital and technical skills – AI/ML, analytics, cloud and hyper-scaler skills to enable scaling through compute and storage elasticity, and IoT sensors.
- 3D and geospatial skills.
- Cyber security skills and awareness as applicable to critical services, like infrastructure digital twins.
- Digital and modern methods of construction skills across sectors, as well as traditional competencies.

8.6. 3D Printing and Additive Manufacturing

This section covers a range of processes and technologies that produce physical objects layer by layer, in an additive manner from a digital 3D model. Industry analysts believe we are only at the beginning of understanding the potential of these technologies. Originally, 3D printing was used for rapid prototyping to fast track and reduce product development costs. But it's now beginning to be used and considered for much broader application.

These technologies allow for new ways of distributed, customisable manufacturing. For example, they have the potential to bring production much closer to its destination and/or directly to the customer. This eliminates the logistics and supply chain costs associated with mass production. And diminished emissions significantly.

The distributed nature of the technology also reduces production risks with smaller batch runs. This will have significant impact on global economies shifting from production and distribution models to more localised on-demand models. An additional plus is to reduce current import and export imbalances between developing and developed nations.

Some of the industries benefiting from, or disrupted by this 3D printing technology include:

- **Medical and dental** – customisation and personalisation to meet medical standards for items such as hip and knee implants, dental crowns, and prosthetics. Although possibly decades away, the technology is being researched and developed for printing skin, bone, tissue, pharmaceuticals, and human organs. (Think 'spray-on skin', cloning, etc.)
- **Aerospace and Defence** – companies such as GE/Morris Technologies, Airbus/EADS, Rolls-Royce, BAE Systems and Boeing, as well as various Defence Departments are using 3-D printing in their research and developments, and current production. Many injectors for jet engines are now 3D printed because they can be made with tighter tolerances than via standard casting and milling approaches.
- **Automotive** – prototyping of high-performance equipment in motor racing through to prototyping for mass-produced vehicles. This is evolving into producing on-demand spare/replacement parts, as opposed to holding inventory. Again, it is being used and further developed in and for the Defence Department, for theatres of war and in Agricultural Departments and industries for spare part printing often only available in other countries and held up by supply chain issues.
- **Jewellery** – reducing the need for high levels of expertise in the manufacturing process including fabrication, mould making, casting, electroplating, forging, silver/goldsmithing, stone cutting, engraving, and polishing.
- **Architecture** – prototyping models of buildings or engineering designs, with some visionary architects looking at direct construction methods.

8.7. Energy Storage-as-a-Service

There has already been significant progress in commercial and household applications for energy storage solutions, and we are likely to see much more. The energy industry is entering yet another industrial revolution with advances in clean energy and storage and distribution technologies. And experiencing the global momentum to bring them into commercial reality.

Renewable generation already exceeds fossil/nuclear generation in various countries. But storage for distributed renewable energy production is still prohibitively capital intensive for most applications. Large scale storage facilities (typically Lithium battery banks today) are currently mainly used for short term grid stabilisation. New technologies such as bulk hydrogen storage would enable broad scale 'Storage-as-a-Service' offerings.

Stationary storage technologies will radically unlock the potential of transient renewable generation technologies like wind and solar. Mobile battery technologies on the other hand, will significantly increase energy density, cycle life, charge and discharge rate, charge/ discharge losses, and self-discharge rate. This will enable a whole raft of new applications with long range EVs just the first. Electric planes and a plethora of devices requiring wired power or fossil fuel engines, will progressively move to long term battery power. Improved battery technology will also radically accelerate the diffusion of the Internet-of-Things (IoT).

e-mobility is expanding, and electric cars, bikes, scooters, and kick bikes are increasingly seen in the streets. Utilisation of all these vehicles will be extended to include energy storage – beyond their initial use as a means of transport. They will go on charge when renewable energy is abundant in the system, and feed energy back into microgrid's battery when needed.

As expected, the 2022 Australian Federal Budget set aside AUD $102 million to establish community solar banks. These will improve access to solar power for those usually unable to access it. Another $224 million will help establish 400 community batteries to reduce fossil fuel reliance. The budget also included $500,000 to be spent next year to develop a future strategy to support offshore renewable projects in Australian waters.

In a related development it is expected that integrated smart grids will become widespread in developed nations, provide the following advantages:

- Smart grids allow energy to be transmitted in both directions through power lines. This means that homes and businesses can add surplus electricity, so no energy goes to waste.
- Smart grids can adapt to the sporadic power output produced by solar and wind energy sources. This feature will make large-scale renewables far more attractive and feasible.
- If a broken transmission line causes a blackout, smart grids' sensors instantly locate the damage and reroute electricity to the area. This will lead to fewer, noticeable blackouts.

In the coming decades, smart grids will expand to even larger scales. Entire regions and countries will begin merging their grids together.

8.8. Quantum Computing

Australia has world-class quantum research capabilities and an emerging quantum technology industry underpinned by the expertise and intellectual property developed in its research institutions.

Quantum computers can solve a variety of problems that are intractable on classical computers. Examples of applications include:

- **Materials** – Simulating quantum mechanical systems is a promising early application of quantum computing. This technique can be applied to fields including chemistry, materials science, and high energy physics.
- **Optimisation** – Quantum computers may potentially find the best solution among varying weighted options more efficiently than classical computers and could be particularly advantageous in vehicle routing and delivery logistics.
- **Machine Learning** – Using quantum systems to train and run machine learning algorithms could solve complex problems more quickly. Potentially improving applications like disease diagnosis, fraud detection, and efficient energy management.
- **Finance** – Quantum computing may have significant advantage over classical computing in computationally intensive finance problems including optimising financial portfolios and analysing their risk.

These examples are far from exhaustive. Researchers around the world are partnering with government and industry to explore potential use cases for the currently available small-to-intermediate scale quantum computers.

Areas and industries that could be revolutionised by quantum computing include:

- **Medicine and materials** – untangling the complexity of molecular and chemical interactions leading to the discovery of new medicines and materials.
- **Supply chain and logistics** – discovering the ideal solution for ultra-efficient logistics and global supply chains.
- **Financial services** – discovering new ways to model financial data and isolate key global risk factors in financial markets.
- **Artificial intelligence** – making aspects of AI such as machine learning more powerful when data sets are very large.

8.9. Hypercritical Datafication

Ubiquitous computing and the Internet-of-Things (IoT) are both huge contributors to the sheer volume of data that's being generated daily. But alongside this machine-generated data, we humans are also generating masses of data through our daily activities. And this shows no signs of slowing down.

Data has indeed evolved exponentially, doubling roughly every 12 months. According to IDC[39] forecasts, by 2025 the global data sphere (defined as the sum of all data created, captured, and replicated on our planet in any given year) is set to reach 163 zettabytes (10 times the 16.1ZB generated in 2016).

By 2025, 20 percent of all data generated is forecasted to be critical to our daily lives and 10 percent hypercritical. Twenty-five percent of that data will be generated in real time, and 95 percent will originate from a hyperconnected IoT world. Cognitive systems (machine learning, natural language processing and artificial intelligence) convert data analysis into intelligence that automates and replicates human actions and judgement. They do so at high speed, mass scale, 'better-than-human' quality, and lower cost.

I predict that, in this decade, governments around the world will recognise and legislate data as an asset.

This is one path to rebalancing the economic, security, privacy and intellectual property opportunity lost in tax revenue by governments and their citizens to internet giants and other organisations due to inadequate data rights, privacy, and protections.

Some countries may also make regulatory and legislative changes for their citizens to regain economic, intellectual property and protection control of their information in their personal, professional, and family lives.

The Consumer Data Right (CDR) introduced into Australian law, aims to help consumers monitor their finances, utilities, and other services. And compare and switch between different offerings more easily. The system also aims to encourage innovation and competition between service providers, making products and services easier to access and better suited to specific needs. The Australian Farm Data Code is an innovative example. It was developed and adopted by the National Farmers' Federation in consultation with industry to promote digital technology by ensuring that farmers are comfortable with the way their data is used, shared, and managed.

These developments are critical as this decade our lives will become exponentially interconnected through more, and more devices being distributed and accessible through the cloud. Given the ability to be predictive through personalisation, self-learning through algorithms, self-provisioning through many forms of biometric identification, and self-adapting to our ecosystems intelligently, autonomously and in real time, their services will become indispensable. All of this and more will be enabled through increased computing capacity, pervasive artificial intelligence, and automation. They will in turn, be enabled by pervasive high-speed connectivity, quantum computing, blockchain, and IoT.

We need to pay close attention to data regulation because much of life in the coming decade will be data driven.

8.10. Artificial Intelligence and Robotics

Our contemporary application of the concept of artificial intelligence has advanced with our knowledge of how the human brain works. AI's earlier use focused on reproducing capabilities such as arithmetic and memory to solve complex calculations. Today it focuses on emulating human decision-making processes and carrying out complex human tasks.

Artificial intelligence has found its place due to the exponential price/performance cost reduction and increase in computer power and storage as detailed in **Section 3**. The tsunami of data has accelerated the focus on AI as a coping mechanism and the past decade or so there have been unprecedented algorithmic advancements. Many of which are associated with machine learning.

The significant breakthrough is machine learning's ability to use algorithms to detect patterns and trends in vast data sets. And then learn how to predict and recommend actions. Deep learning is a subset of machine learning but can process a broader range of data sets. Deep learning has interconnected layers of algorithm-based calculators called neurons (a bit like our brain), that form a neural network by feeding data into one another.

Artificial intelligence is widely embraced in consumer markets. Natural language-based virtual assistants such as the Google Assistant, Microsoft Cortana, and Apple's Siri are now widely used through our smartphones, smart vehicles and in our smart homes. Commercial applications are also rapidly emerging. For example, interpreting video feeds from drones that are carrying out visual inspections and – in the case of the floods and Surf Life Saving Australia[40] – rescuing lives.

AI is one of the most exciting fields in robotics and a significant interdisciplinary area of engineering research. This includes mechanical construction, electrical components and computer programming and human interface technologies.

A tidal wave of invention is also occurring in robotics, enabled by inexpensive sensors and control systems. Robots have moved out of traditional industrial manufacturing, assembly, and painting jobs, and into hospitals, roads, aerial vehicles, small businesses, and the home. Robots can already see, hear, feel, smell, touch, manipulate objects, walk, run, jump, dance, talk, and do both macro-scale and nano-scale work.

Breakthroughs in material science, locomotion and sensory technologies have improved the orchestration of networked machines and their ability to move in different environments. Cognition enables them to be choreographed.

Artificial intelligence allows machines to perform complex navigation and interactivity in an autonomous way.

The combination of artificial intelligence and robotics raises many complex issues in areas such as responsibility and accountability for actions or inactions and ethical standards. For example, unintended programmatic bias, data regulation and mechanisms for resolving conflict between autonomous and non-autonomous actors in the systems.

The automotive industry has been one of the largest users of automated robots in assembly lines. All industries will be impacted by automation over the coming years, and many parts of the economy will be boosted in terms of improved productivity, elimination of labour-based jobs and the creation of new jobs and services.

The manufacturing, transport, and logistics industries are likely to be the sectors that will continue investing to further improve production, assembly, and distribution. But other sectors are rapidly adopting these technologies including:

- **Manufacturing** – assembly lines e.g., the automotive industry.
- **Mining** – extracting and processing minerals from the ground.
- **Healthcare** – surgical procedures.
- **Agriculture** – weed and pesticide management.
- **Entertainment** – e.g., Gaming etc and movies e.g., Sci-Fi.

Exhibit 29: Case Study: Australian farmers invest in robots, autonomous vehicles to combat worker shortage[41]

Robots are being used to commercially farm more than 405,000 hectares of Australian farmland, with growers seeing them as the answer to workforce shortages.

Growers there are investing in autonomous vehicles that can be programmed to slash and spray herbicides more efficiently.

Labour shortages and extreme weather have accelerated the company's investment in the vehicles.

Robots can be programmed to slash and spray herbicides more efficiently. Growers are seeing environmental and productivity benefits in investing in robots and new jobs are expected to be created in supporting industries.

8.11. Blockchain/Distributed Ledger

As industries, supply chains and ecosystems become programmatic, that is programmed into software – services will become automated. They will be replaced by decentralised autonomous organisations (DAO), also known as decentralised autonomous corporations (DAC).

A DAO is a network that runs itself. It's able to operate autonomously due to the clever use of mechanisms that align economic incentives (distribute risks/rewards) using software. The network is structured and incentivised to operate without the use of trust, third parties or centralised powers. In a financial services context, finance becomes 'programmatic'. Distributed ledger technology, aka blockchain, restructures value chains by disintermediating (removing intermediaries), disaggregating (separating processes into components) and dematerialising (shifting from physical to digital form).

The disruptive impact of a DAO is contained within a property that is often overlooked. It operates in real time (see **Exhibit 30**).

Exhibit 30: Case Study: DOA Based Audits

For example, 'DOA based audits' would no longer become a 'point-in-time' retrospective analysis (e.g., auditing last year's accounts). Audits occur continuously and in real time. There is significant interest in this model in government as a primary regulatory innovation.

If a regulator knows as soon as there is variation it can potentially do away with the whole cycle of 'record, audit, and potentially a retrospective fine (and often protracted litigation)', to a mode where regulators intervene with corrective action as soon as variation occurs.

Imagine the disruptive impact this would have on the legal, accounting, and professional services across regulated markets. Think financial services, healthcare, energy etc. Importantly, imagine the transformative impact this could have on first, second and third lines of defence in risk management.

Areas and industries that could be revolutionised by technologies such as blockchain and distributed ledger include:

- **Crypto-currencies** – one of the best-known applications e.g., Bitcoin, is described as a decentralised virtual currency.
- **Supply chain and logistics** – self executing contracts that transfer risk and value and insurance, verifying the transaction or any associated claims with pre and post-authorisations and instructions.
- **Currency exchange and remittance** – perhaps the most advanced use cases with financial institutions focused on cross border payments. Pioneers include Ripple.
- **Provenance and chain of custody** – time stamping features for agriculture (e.g., from paddock to plate), and for precious stones e.g. diamonds.
- **Record management** – improve medical record management to store healthcare data and manage medical records.
- **Decentralised markets** – enabling people anywhere to trade with one another without requiring an institution.

8.12. Climate Tech innovation

Climate technology analysts[42] define Climate Tech as a scalable tech-enabled solution for climate change by:

- Directly reducing, avoiding, or removing emissions. Or enabling others to do so.
- Directly mitigating the effects of, or adapting to a changing climate, or enabling others to do so.

Partnerships with academia and start-ups can help accelerate innovation. The new wave of innovation is powered by carbon-intelligent platforms, with plug and play sensors, data integration from multiple sources, and visualisation and simulation of emissions for optimisation.

Australia and New Zealand have developed a broad, diverse, and thriving Climate Tech industry that is making a significant contribution to our economy. It is doing so through investment, employment, offshore development, creation of intellectual property and much more (see **Exhibit 31**). Some of the rapidly growing parts of the Climate Tech sector are in industries critical to Australia's carbon reducing goals such as agriculture and food, the circular economy, renewables and much more.

Exhibit 31: Profile of Australian and New Zealand Climate Tech Sector

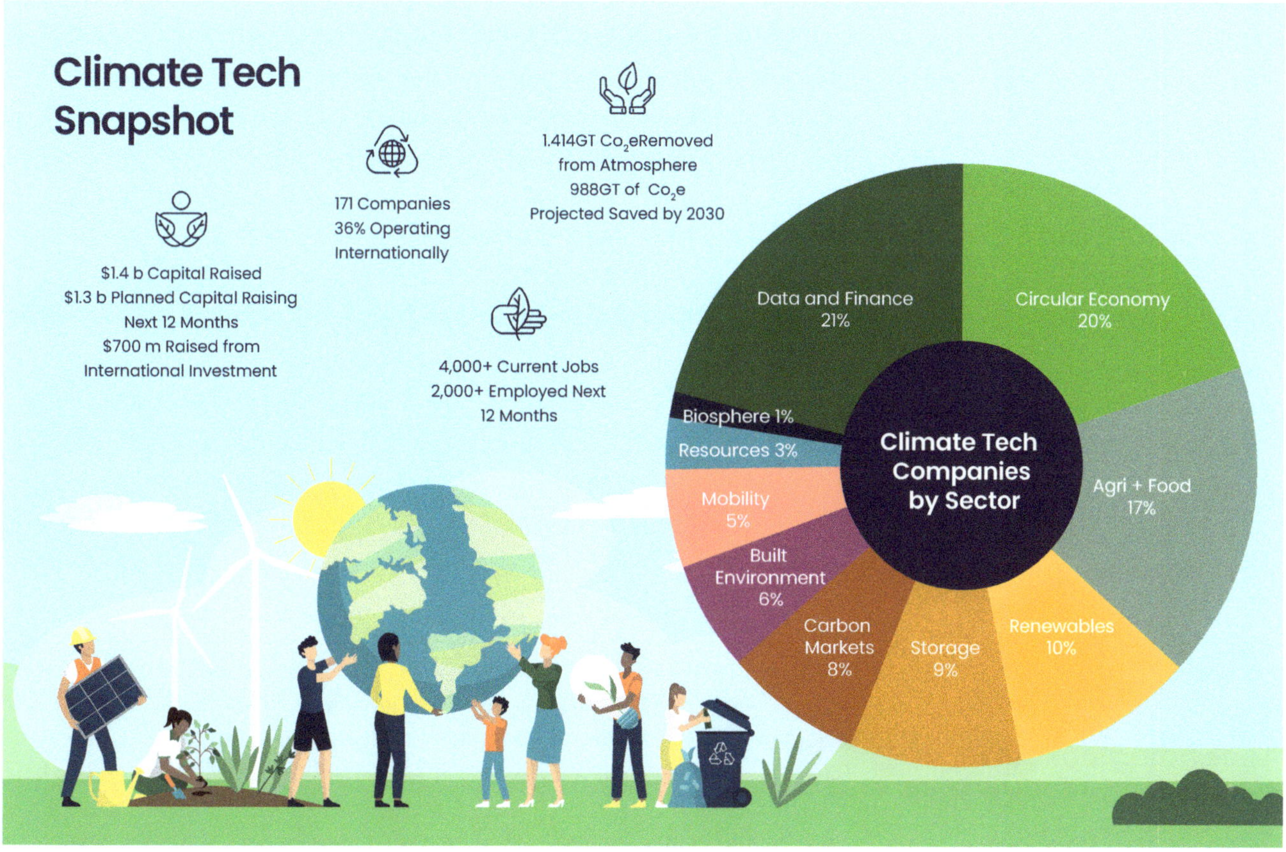

Data Source: Climate Salad

8.13. The Circular Economy

To keep volumes of major commodities such as metals aligned with economic growth, we will need to embrace circular models more quickly. This means sourcing much less from virgin materials, using recycled content and remanufactured products, and generally rethinking the material economy. Water is and will be a stressed resource. It is likely that many cities will be in a constant state of water shortage.

We will also need to change how we produce new materials. Much new materials engineering is now effectively done in software rather than the lab. We are reaching the point where software can take a set of requirements and constraints and design a material to meet them. For example, this is how a lot of exploration for new high temperature superconductors and some classes of pharmaceuticals is done. Work is well underway on the next phase of this journey, where the software also designs the production process for the material.

New materials are emerging that are lighter, stronger, recyclable, conductive, transparent, and adaptive. For example:

- Bio-inspired plastic that is light enough to permit flight and thin enough to accommodate flexibility.
- Ultrathin platinum-hydrogen fuel cell vehicles that can provide clean transportation in the future.
- Graphene, with its strong, conductive, flexible properties, which can be applied to next generation electronic devices, or even sewn into our clothing.
- Stanine, which was created as a topological insulator that scientists say is the natural successor to copper interconnects in computers.

Exhibit 32: Case Study: Decarbonising Australia's agriculture, steel production and other heavy industries.

A new $90 million research initiative was recently launched involving some 30 companies including the CSIRO to develop and commercialise new products.[43]

The CSIRO was reported to have contributed AUD$22 million 'Towards Net Zero Mission' with the balance coming from external partners and investors. These include aircraft manufacturer Boeing and Qantas that are looking into synthetic aviation fuel and hydrogen fuels.

Other collaborators include Australia's largest steel producers that seek to replace coal in steelmaking for other materials that will help iron ore change its chemical composition in a blast furnace.

In a first for the $90 million program, the CSIRO's FutureFeed initiative of a seaweed-based feedstock that reduces methane emissions from livestock by more than 80 percent, is now on the market.

8.14. Technological and Scientific Development – Closing Point

In closing this section, it is indeed our optimism about technological and scientific advancement, that is an intrinsic quality in Australia's pioneering culture. That will be the secret to addressing the many climate change challenges ahead.

Participants in this *Australia 2030 Research*© were invited to qualitatively share their thoughts on why they were either positive or negative about the role of technological and scientific developments to support climate change over the coming decade (see **Exhibit 33**). Pleasingly they were overwhelmingly positive.

These are their words creatively illustrated in the word cloud.

Exhibit 33: When you think about your life over the coming decade, are you positive or negative about the role of technological and scientific developments to support climate change? (Select 1 Response) (optional why or why not?)

Source: *Australia 2030 Research*© Rocky Scopelliti (n=61)

9.
Conclusion

The blueprint for how Australian organisations can increase their capacity to adapt to a world in accelerated climate change.

This *Australia 2030 Research*© has identified the risk of 'exponential lag' if Australian organisations are to meet the 2030 carbon emissions target. That lag is the gap between the time when investment decisions are made, transformations implemented and the time when decarbonisation benefits occur. Net zero carbon emissions awaits Australia at the end of the exponential curve of change – not the linear curve.

The materialisation of that risk in 2030, will further accentuate an organisations' capacity to reset onto the necessary trajectory to reach their net zero target by 2050. For Australia, that exposes investment risks for the forecasted $20 trillion of capital required by 2050, to which will need to be invested differently, and be structurally different, to compete in a low emissions world.

The major decision-making barrier board directors and chief executives currently face, is lack of clarity around the benefits and ROI of their environmental programs. This is most likely stifling decision making on investment decisions. However, without question, the timing for those investment choices is now.

This situation is at odds with the board director's dilemma. How to adapt toward the new low carbon emissions world, when their organisations are today, generating good returns. Exacerbated by the facts that the transformation costs are upfront, and the dividends back end.

That dilemma was clearly evidenced by the facts of the *Australia 2030 Research*©.

The leaders climate change 'To Do' Checklist includes:

1. **Measures to establish:** Many organisations are yet to establish their environmental strategies, policies, and programs.
2. **Investment to undertake:** This leaves organisations exposed by under investment in the enabling technologies before they can further benefit from emerging technologies.
3. **Emissions baseline to establish:** Organisations need to urgently commence and deploy measures across their organisations to establish a carbon emissions baseline.
4. **Decision integration to occur:** Those measures will enable decision making to become more integrated across strategic and execution considerations.
5. **Shift to strategy-driven:** Importantly, it will further assist organisations evolve their environmental goal-setting maturity from principally being Compliance and Risk Driven, to Strategy-Driven.
6. **Stakeholder reputation to uphold:** With that development, organisations will then be much better equipped to influence all stakeholder groups with their investments in their environmental initiatives that today, are mostly concentrated on customers and increasing the company's brand and reputation.

Given environmental programs are both data and technology intensive with highly fragmented and diverse sources and require major organisational transformations, Chief Information Officers/Technology Leaders are the executives with the necessary leadership and skills to navigate the complexities. They will be able to orchestrate such a transformation of this scale.

Their responsibilities will have to expand two-fold. Vertically, creating Systems Leadership outside the organisation. And within the organisation, becoming the Environmental Enablers horizontally.

The prize for this systems leadership-driven approach could see an economic gain for net-zero transformation in the order of $890 billion added to Australia's GDP by 2070.

For Australia, the optimism amongst those in younger demographics who will be tasked to carry the batton to 2050, about the speed of technological and scientific change required to support Australia's environmental goals over the coming decade, provides a cultural advantage that could deliver a dividend as Australia transitions into a low carbon economy.

Australia has a remarkable history of innovation. We need to remind ourselves how important it is to both look to and trust the future. With the next generation of high value-adding emerging industries, Australia has a real opportunity. As an advanced economy it can and must invest in the high value-added pre and post-production stages of existing and emerging industries to create the new economic value.

Acknowledgments

This report was made possible by the generous contributions of Australian board directors and chief executives and professionals who not only contributed to this *Australia 2030 Research*©, but also provided me with encouragement, advice, and inspiration for this publication. I am sincerely humbled, grateful, and appreciative to you all.

Glossary

Term	Definition
Additive Manufacturing	Additive manufacturing is defined as a material joining process, whereby a product can be directly fabricated from its 3D model, usually layer upon layer.
AgTech	Agricultural technology or agrotechnology (abbreviated variously as agtech, agritech, AgriTech, or agrotech) is the use of technology in agriculture, horticulture, and aquaculture with the aim of improving yield, efficiency, and profitability.
Artificial Intelligence	Artificial intelligence (AI) is the ability of a computer program or a machine to think and learn.
Augmented Reality	Augmented reality (AR) is an interactive experience that combines the real world and computer-generated content.
Autonomous Vehicles	Autonomous vehicles are based on an automated driving system refers to a vehicle system that uses both hardware and software to exercise dynamic control of a vehicle on a sustained basis.
Bio Refineries	A biorefinery is a refinery that converts biomass to energy and other beneficial byproducts (such as chemicals).
Biotechnologies	Biotechnology is the integration of natural sciences and engineering sciences to achieve the application of organisms, cells, parts thereof and molecular analogues for products and services.
Blockchain	A blockchain (or block chain) is a method of storing a list of entries, which cannot be changed easily after they are created.
Cellular Agriculture	Cellular agriculture refers to the development of agricultural products – especially animal products from cell cultures rather than the bodies of living organisms.
Circular Economy	A circular economy (also referred to as circularity and CE) is a model of production and consumption, which involves sharing, leasing, reusing, repairing, refurbishing, and recycling existing materials and products for as long as possible.
ClimateTech	ClimateTech (also known as clean technology, in short CleanTech) is any process, product, or service that reduces negative environmental impacts through significant energy efficiencies
Cobot	A cobot, or collaborative robot, is a robot intended for direct human robot interaction within a shared space,
Digital Twin	A digital twin is a virtual representation of a real-world physical system or product (the physical twin) that serves as the indistinguishable digital counterpart of it for practical purposes, such as system simulation, integration, testing, monitoring, or maintenance.
Distributed Ledger Technology	A distributed ledger (also called a shared ledger or distributed ledger technology or DLT) is the consensus of replicated, shared, and synchronised digital data that is geographically spread (distributed) across many sites, countries, or institutions.
Distributed Autonomous Organisation	A Distributed or Decentralised Autonomous Organisation is an organisation represented by rules encoded as a transparent computer program, controlled by the organisation members, and not influenced by a central government.
Emerging Technologies	Emerging technologies are technologies whose development, practical applications, or both are still largely unrealised. These technologies are generally new but also include older technologies finding new applications. Emerging technologies are often perceived as capable of changing the status quo.

Term	Definition
Enterprise Technologies	Enterprise technology is the information technology (IT) tool that businesses use for efficient production operations and back-office support. These IT tools cover database management, customer relationship management, supply chain management, business process management and so on.
Genomics	Genomics is an interdisciplinary field of biology focusing on the structure, function, evolution, mapping, and editing of genomes. A genome is an organism's complete set of DNA, including all of its genes as well as its hierarchical, three-dimensional structural configuration.
Gene Editing	Gene editing is an emerging molecular biology technique that makes very specific targeted changes by the insertion, deletion, or substitution of genetic material in an organism's DNA to obtain desired results.
Geoengineering	Climate engineering, (or geoengineering) is defined as deliberate and large-scale intervention in the earth's climate system to reduce/reverse climate change or, more broadly, the use of technology to modify and manipulate the climate.
Green IT	Green computing, green IT, or ICT sustainability, is the study and practice of environmentally sustainable computing or IT.
Greenwashing	Greenwashing (a compound word modelled on "whitewash"), also called "green sheen", is a form of advertising or marketing spin in which green PR and green marketing are deceptively used to persuade the public that an organisation's products, aims, and policies are environmentally friendly.
Holographic Display	A holographic display is a type of 3D display that utilises light diffraction to display a three-dimensional image to the viewer.
Internet-of-Things	The Internet of Things (IoT) describes physical objects (or groups of such objects) with sensors, processing ability, software, and other technologies that connect and exchange data with other devices and systems over the Internet or other communications networks.
Light Energy	Light is a form of electromagnetic radiation that shows properties of both waves and particles. It is a form of energy. Light also keeps the Earth warm. Light exists in tiny energy packets called photons. Each wave has a wavelength or frequency.
Material Science	Materials science includes those parts of chemistry, physics, geology, and biology that deal with the physical, chemical, or biological properties of materials. It is usually considered an applied science in which the properties it studies are used industrially.
Microgrids	A microgrid is a local electrical grid with defined electrical boundaries, acting as a single and controllable entity. It can operate in both grid-connected and in island mode. A 'stand-alone microgrid' or 'isolated microgrid' only operates off-the-grid and cannot be connected to a wider electric power system.
Mixed Reality	Mixed reality (MR) is a term used to describe the merging of a real-world environment and a computer-generated one. Physical and virtual objects may co-exist in mixed reality environments and interact in real time. Mixed reality is largely synonymous with augmented reality.
Motion Energy	In physics, the kinetic energy of an object is the energy that it possesses due to its motion. It is defined as the work needed to accelerate a body of a given mass from rest to its stated velocity. Having gained this energy during its acceleration, the body maintains this kinetic energy unless its speed changes.
Nanotechnology	Nanotechnology, also shortened to nanotech, is the use of matter on an atomic, molecular, and supramolecular scale for industrial purposes.

Term	Definition
Quantum Computing	Quantum computing is a type of computation whose operations can harness the phenomena of quantum mechanics, such as superposition, interference, and entanglement. Devices that perform quantum computations are known as quantum computers.
Robotic Process Automation	Robotic process automation (RPA) is a form of business process automation technology based on metaphorical software robots (bots) or on artificial intelligence (AI)/digital workers. It is sometimes referred to as software robotics (not to be confused with robot software).
Smart Grid	A Smart Grid is an electricity network that can cost efficiently integrate the behaviour and actions of all users connected to it – generators, consumers and those that do both – to ensure an economically efficient, sustainable power system with low losses and high levels of quality and security of supply and safety. A smart grid employs innovative products and services together with intelligent monitoring, control, communication, and self-healing technologies
Synthetic Biology	Synthetic biology (SynBio) is a multidisciplinary area of research that seeks to create new biological parts, devices, and systems, or to redesign systems that are already found in nature.
Unmanned Ariel Vehicles	An unmanned aerial vehicle (UAV), commonly known as a drone, is an aircraft without any human pilot, crew, or passengers on board. UAVs are a component of an unmanned aircraft system (UAS), which includes adding a ground-based controller and a system of communications with the UAV.
Vertical Farming	Vertical farming is the practice of growing crops in vertically stacked layers. It often incorporates controlled-environment agriculture, which aims to optimize plant growth, and soilless farming techniques such as hydroponics, aquaponics, and aeroponics.
Virtual Reality	Virtual reality (VR) is a simulated experience that employs pose tracking and 3D near-eye displays to give the user an immersive feel of a virtual world.
3D Printing	3D printing or additive manufacturing is the construction of a three-dimensional object from a CAD model or a digital 3D model.
5th Generation Mobile Networks	In telecommunications, 5G is the fifth-generation technology standard for broadband cellular networks

Source: Wikipedia

Endnotes

1 Media Release, (2022), Prime Minister for Climate Change and Energy, Assistant Minister for Climate Change and Energy. 'Stronger Action on Climate Change'

2 National Australia Bank, (July 2022), (https://news.nab.com.au/news/opportunity-in-low-emissions-world.)

3 McKinsey Sustainability, (April 2022), 'Delivering the climate technologies needed for net zero'

4 Named after Gordon Moore the co-founder of the computer chip company Intel

5 Ray Kurzweil, (1999), 'The Age of Spiritual Machines' (Viking Press)

6 Climate Change Initiative, (Accessed in September 2022), https://climate-governance.org/

7 AICD, (2022), Accessed on 9 October 2022; https://www.aicd.com.au/tools-and-resources/climate-change/climate-governance-forum-2022-highlights.html

8 Deborah Tarrant, (2022), AICD 'Beware the risk of greenwashing'

9 Peter Hannam, (2022), The Guardian, 'ACCC begins 'greenwashing crackdown on companies' false environmental claims'

10 CRN, (2022), 'ACCC warns about green washing'

11 AFR, (2022, Hannah Wootton, 'Energy company fined $53,000 in first ever green washing penalty'.

12 McKinsey Technology, (September 2022), The green IT revolution: A blueprint for CIOs to combat climate change'

13 Deloitte Access Economics (2022), Clean Cloud: Meeting the climate challenge in Australia & New Zealand. Available at: https://services.google.com/ fh/files/misc/deloitte_clean_cloud_meeting_the_climate_ challenge_aunz_report.pdf

14 Telstra (2017), Connecting with the cloud. Available at: https://insight.telstra.com.au/content/dam/insight/TelstraCCE_Report_24pp_FA_web_spreads-09.06.pdf

15 McKinsey Technology, (September 2022), The green IT revolution: A blueprint for CIOs to combat climate change'

16 United Nations, (2015), 'World Population Aging'

17 Capgemini and LinkedIn, (2022), 'The Digital Talent Gap – Are companies doing enough?'

18 World Economic Forum, (2022), 'Why are young people not prepared for the jobs of the future'

19 Australian Government The Treasury, (2022), 'Jobs + Skills Issues Paper'

20 Deloitte Access Economics (2021), ACS Australia's Digital Pulse. Available at: https://www2.deloitte.com/ content/dam/Deloitte/au/Documents/Economics/deloitteau-economics-australias-digital-pulse-2021-160621.pdf

21 RMIT and Deloitte Access Economics, (2021), 'Ready, set, Upskill: Effective Training for the Jobs of Tomorrow' https://assets.online.rmit.edu.au/cdn/ff/DdDm2DW8b5WnLbn2dQy1fgOenjzqCrpzYQv8F-vyngl/1612754024/public/whitepapers/RMIT percent20Online_ReadySetUpskill.pdf

22 Department of Education, Skills, and Employment (2019), "uCube". Available at: https://highereducationstatistics.education.gov.au/

23 Climate Council, (2022), 'Power Up: Ten Climate Gamechangers'

24 World Economic Forum, (2022), 'Digital solutions can reduce global emissions by up to 20 percent. Here's how

25 Optus Media release, (2020), New ANU-Optus partnership to build national defence system against catastrophic bushfires

26 One single life year is equivalent to an individual losing one full year of health or lost life.

27 Google, (2022), 'Google's Economic Impact in Australia: Helping build Australia's digital future through technology'

28 AIIA, (2021), 'Growing Globally Competitive Industries – Powered by Australia's innovative technology'.

29 NSW Government, (https://invest.nsw.gov.au/sector-opportunities/agtech)

30 https://www.ausagritech.org/

31 Innovation Aus.com, (2022), Jennifer O'Brien, Sustainable, reliable food production with modern, vertical farming'

32 https://freopp.org/australia-health-system-profile-11-in-the-world-index-of-healthcare-innovation-164225723e08

33 https://nationalindustryinsights.aisc.net.au/industries/health

34 https://www.industry.gov.au/publications/australian-manufacturing-performance-report

35 https://www.industry.gov.au/data-and-publications/make-it-happen-the-australian-governments-modern-manufacturing-strategy/our-modern-manufacturing-strategy

36 AIIA Media Release, (2022), 'AIIA calls for joined-up digital approach to disaster preparedness: Increasing frequency of bushfires and floods requires national approach'.

37 https://www.spatial.nsw.gov.au/what_we_do/projects/digital_twin

38 World leading Spatial Digital Twin launched in NSW | NSW Government46

39 International Data Corporation (IDC) is the premier global provider of market intelligence, advisory services, and events for the information technology, telecommunications, and consumer technology markets.

40 In Australia, lifeguards used a drone to save two people from drowning at Lennox Head in New South Wales. The drone located the swimmers in trouble and dropped lifesaving floatation equipment to them until rescuers arrived. This was all done within 70 seconds from when the drone was launched.

41 ABC News, Johanna Marie (2022), 'Australian farmers invest in robots, autonomous vehicles to combat skills shortages'

42 Climate Salad, (2022), '2022 Australia & New Zealand Climate Tech Industry Report'

www.ingramcontent.com/pod-product-compliance
Lightning Source LLC
LaVergne TN
LVHW072119210926
848482LV00027B/706
* 9 7 8 0 6 4 5 6 6 3 0 3 7 *